Why Can't We All Just Get Along?

HOW SCIENCE CAN ENABLE
A MORE COOPERATIVE FUTURE

Christopher Fry

Henry Lieberman

Dedication

To Democracy: The ideal of eliminating inequality is laudable. To further it, we now need something fairer than voting.

To Communism: At least you had the idea that people should cooperate, but forced centralization corrupted you beyond repair.

To Capitalism: We appreciate you getting us this far. Motivation by wealth is powerful. But that power corrupts. Distributing the means of production will help.

To Socialism: Reducing inequality helps, but there's a better way than just taxing the rich.

To Teachers: We appreciate your dedication to students. That shouldn't be undermined by the Common Core and Every Child Left Behind. Please work for Constructionism and intrinsic motivation.

To Parents: Teach your children well. Their future depends more on their ability to cooperate than their ability to compete.

To The Unemployed: We'll all join you soon enough. Meantime, learn how to make.

To Makers: Please continue to innovate and share. Civilization depends on it.

Acknowledgments

We are indebted to the MIT Media Lab, the MIT Computer Science and Artificial Intelligence Lab, and Harvard University for providing the fertile grounds for numerous water cooler conversations that helped crystallized the complex concepts of this work.

Our core concepts were first presented at two provocative conferences in LA in June 2015:

Seizing an Alternative: Toward an Ecological Civilization, at Claremont College

Limits 2015: Workshop on Computing within Limits at UC Irvine

where the feedback helped us improve our presentation of these ideas immeasurably.

We'd also like to thank John Werner, Mark Connell, Irma Rastegayeva, and Erin Rubin for the invitation to present a TEDx Beacon Street talk in November 2017. Watch it on http://www.tedxbeaconstreet.com/videos/whycantwe/

Interviews with your authors at https://tedxbeaconstreet.com/videos/interview-with-christopher-fry/ and https://tedxbeaconstreet.com/videos/interview-with-henry-lieberman/

We thank the participants in our Spring 2018 MIT course whose textbook you hold in your hands, especially Dana Bullister, David Lewit, Howie Goodell, James Wiggleworth, Nathan Kaiser, Patrick Breslin, Richard and Toby Shyduroff, Suzanne Watzman, and Valeria Staneva.

Colleagues and friends have reviewed these chapters to hone them into a coherent whole. We especially appreciate the careful reading and insightful comments of:

Bonnie Nardi, Candy Leonard, César Hidalgo, Cindy Mason, David Lewit, David Ungar, Fred Lakin, Jamie Macbeth, Jessica Artiles, Joscha Bach, Karthik Dinakar, Ken Kahn, Kevin Kane, Louis Smith, Maria Karam, Mark S. Miller, Ray Garcia, Susan McLucas, Suzanne Watzman, Ted Selker, and Yen-Ling Kuo.

Thanks also to the following people for helpful feedback and conversations around the issues presented here.

Abdullah Almaatouq, Adam Jaroszweiski, Sheng-Ying (Aithne) Pao, Alex Faaborg, Alex Wisner-Gross, Andrew Gordon, Ben Sherman, Bill Tomlinson, Birago Jones, Bob Mohl, Carol Adams, Catherine Kreatsoulas, Catherine d'Ignazio, Cathy Fry, Chris Schmandt, Dale Joachim, Dennis Peterson, Earl Wagner, Edward Shen, Elaine Raybourn, Elizabeth Rosenzweig, Emil Jacobs, Eric Drexler, Ethan Zuckerman, Fox Harrell, Gene Nishinaga, George Mokray, Gerhard Weikum, Glenn Iba, Gloria Brown-Simmons, Harry Halpin, Heidi Roper, Herve Boillot, Ivan Sysoev, Iyad Rahwan, James Graham, James Redd, Jeremy Fry, Josh Tanenbaum, Joy Adowaa Buolamwini, Judy Sacknow, Kent Gilson, Kristin Hall, Larry Susskind, Lea Peersman-Cherigie Pujol, Leila Moy, Lorenzo Coviello, Louie Weitzman, Manushaqe Muco, Mark Van Harmelen, Marvin Minsky, Matt Weiss, Matthew Nock, Nathan Mathias, Nick dePalma, Pattie Maes, Paul Cotran, Pedro Cuellar, Philippe Piernot, Richard Stallman, Robert Gerke, Roger Levy, Scott Greenwald, Sumner Silverman, Susan McLucas, Tom Moy, and Xiao Xiao.

Thanks to Walt Lieberman and to Cathleen Schaad for the cover, illustrations, and help with preparation for publication.

Contents

Introduction

In 1991, African-American Rodney King was stopped by white police officers, for speeding. Angry at King for leading them into a car chase, the police brutally beat King while a bystander captured the scene on video. After the officers were investigated and acquitted despite the video evidence, violent riots erupted in the African-American neighborhoods of Los Angeles as a reaction to the perceived racism and police brutality. In 2014-5, similar events transpired in Ferguson, Missouri, New York City, Baltimore and other American cities.

Exasperated at both the needless brutality he suffered, and at the damage suffered in the violent reaction to the incident, King summed up his feelings in a simple, poignant question:

"Why can't we all just get along?" [1]

Indeed, why can't we?

We relate this story not to make a point about racism and police brutality, though those are very serious problems.

Think about it. Here was a guy who had every reason to be bitterly angry about many things. Instead, his Zen-like reaction was to step back from the details of the situation and ask: Isn't it obvious that we should be able to do better?

That's what we're asking you, Dear Reader, to do. Step back from the details of today's frightening headlines, and ask: *Do we really need to have a world of war, poverty, and hatred? Isn't it obvious we should be able to do better?*

From playground bullying, to wars involving millions of deaths and lasting for decades, the world is full of violence that most reasonable people recognize is pointless. People in governments seem more interested in

---

1   Actually, King didn't use these exact words; his rambling speech can be seen in [King 1992]. But that was the intent. And that was how he was quoted in the widespread discussions that followed.

obtaining and maintaining power rather than meeting the needs of their constituents. Corporations spend billions of dollars making products that provide poor value for customers. They fight with consumers and other companies through deceptive advertising. They exploit their workers and corrupt governments. Meanwhile, peoples' real needs go unmet.

There must be some better ways of organizing society. Neither traditional Capitalism nor Communism, Authoritarianism nor Democracy seem to hold the keys for avoiding these kinds of problems. Are these problems innate? Does human nature have some kind of inborn need for aggression and stupidity that won't ever go away?

We don't think so. Basically, we can all just get along. Maybe not every single time, maybe not perfectly. But rethinking how we organize ourselves and how we make decisions could result in avoiding large-scale, human-caused disasters that threaten humanity. For the first time in history, we have technology that could help us create wealth, distribute it equitably, and deal with the complexity of difficult systems-level problems. Resources, human and material, that now go to wasteful activities could be channeled to improve life for everybody.

First, we have to understand the temptations and traps that lead people to act against the best interests of themselves and others. Undoubtedly, some criminal and antisocial activities are committed by people who have evil intentions, lack moral character, or who have mental illnesses that cause them to be sociopathic. But evil and sick people only make up a few percent of the total population.

Large-scale problems tend to occur because relatively "normal" people can find themselves "sucked in" to behavior that winds up being bad for everyone as a whole. They find themselves (or perceive themselves to be) in desperate situations; they lose their temper; they get greedy; they are oblivious to the impact of their actions on others; they work for organizations that have perverse incentives. It is by reducing the occurrence and impact of these situations that we can get on a better path.

We see many reasons to be hopeful in today's world. One is mathematical. In this book, we will show that many of these situations can be described by a simple mathematical model, the Prisoner's Dilemma, the basics of which can be easily understood at the high-school level.

What this mathematical model does is explain the tradeoff between cooperation and competition. Neither is best under every circumstance. But there are many situations where it may seem like competition is best, but, in fact, cooperation would be better if everybody understood the general case. This is important. It can get us out of arguing about who's responsible, who's at fault, and who's to blame. We can shift to the more productive question: how do we break out of an unproductive pattern?

Another reason to be hopeful, is computer technology. Whether competition or cooperation is best depends on a few factors that may tip the math in either direction. Information technology is now changing these factors so that many situations that used to favor competition are now favoring cooperation. But our society is still stuck in old competitive patterns.

Technology alone cannot fully solve people problems, but in many cases, it can help relieve stresses that tax people's limited ability to perceive, understand, reason, decide, communicate, and produce.

It's our contention that two root causes of a lot of these problems are *scarcity* and *complexity*. When there's scarcity, people get desperate, and desperate people are prone to do stupid or hostile things. When there's too much complexity, people get overwhelmed and fearful, again tempting them to do stupid or hostile things. Robotics and 3D printing can make us more productive, relieving scarcity. Artificial intelligence, end-user programming, and computer-assisted collaborative software can help us better manage complexity.

Finally, although it may not seem like it, we are indeed making progress in psychology and social relations. We are improving people's ability to get along with one another by understanding patterns of communication between people, helping people better manage their emotional reactions to situations, and inventing better ways of organizing social cooperation. In short, we're getting better at understanding ourselves.

Some of our solutions have been discovered and demonstrated on a small scale, but are not yet mainstream or commonplace. Related work has taken place in fields like psychology, education, conflict resolution, negotiation, mediation, cooperative enterprise and small-scale intentional communities. We connect the dots.

Both optimists and pessimists can find plenty of evidence in today's world to support their respective views. We confess right away: we're optimists. If you're a hardcore pessimist, we won't be able to convince you.

The rest of you might be inclined towards optimism, but you may have trouble seeing how we even have the possibility of winning, given all of today's problems. Hear us out. We'll explain how technological changes in society hold the promise of enabling a much more positive future. Seeing the possibility for such change encourages the optimism we'll need for taking advantage of exactly these opportunities. By weaving together a tapestry of new interdisciplinary threads, maybe we really can all just get along.

What are you getting yourself into?

You are getting yourself into an outrageously ambitious project: solving the world's major social problems.

Our strategy starts with bringing you on an exploration that includes identifying our most serious problems, analyzing underlying causes, recognizing commonalities, and providing solutions that address root causes.

We hope you'll find insightful solutions that are, at least unconventional, perhaps even innovative. This means they are, by and large, not proven. If you said we're skating on thin ice, you'd be correct. (And, with global warming, the ice we're all skating on will be getting thinner.)

We're going to lace up anyway. The temporary band-aids of conventional wisdom are not working. Band-aids may simply fill a niche precluding real solutions. All that's necessary for the bad guys to win is for the good guys to be distracted with the insufficient.

We've tried to write in a fun and entertaining style, even when we're talking about dead serious issues. We're trying to get you to reflect on the absurdity of some the contradictions of contemporary life. We hope it'll be obvious which parts are intended to be tongue-in-cheek. Unlike comedians, though, we present possible solutions.

Please forgive us, if sometimes we rant against the stupidity we see in the world. It's not so much anger, as it is righteous indignation at seeing problems going unsolved, and people suffering needlessly. It was good therapy for us to get it off our chests, and we hope that reading it will help you let off some steam, too. But it will undoubtedly upset some readers.

Don't start nitpicking every sentence until you grasp the overall argument. If you find yourself unable to continue reading because too many objections pop into your head, go directly to the FAQ, where we answer the most common objections.

One of the reasons our problems are so resistant to fixes is that restructuring any one of these institutions would merely result in the others restoring it to status quo. To at least some extent, its necessary to solve all the big problems before solving any of them.

Tough? Sure. But put yourself in the shoes of someone from Ferguson or Syria. Or your children.

A journey through solution-space

This book is structured in 5 parts, each of which is an inquiry into the question of whether, and how, it is possible to achieve an aspect of cooperation. You don't have to read the book in order, cover-to-cover. Feel free to skip around to the parts that interest you most. Keep the book in your bathroom and read a bit of it at a time.

The first part, *What keeps us from getting along?* presents our fundamental argument, drawing from mathematics, psychology, and evolutionary theory. It gives you the tools for thinking about the rest of the issues we will cover. We do have to explain some scientific concepts, but we tried to make sure it's easily understandable to everybody. If you have a little trouble with it, don't let it stop you—skip it and read later sections of the book. When you've read more of the book, come back to it and you might then find it more approachable.

The second part, *Does human nature allow us to get along?* debunks arguments that say that aggression, conflict, greed, and war are inevitable. This stance of inevitability is perhaps the biggest obstacle we have, since if you believe something is inevitable, you won't be motivated to change it. Sure, there's plenty of history that might make you pessimistic. But we claim that these negative tendencies are due to scarcity and societal conditions. We now have the technology to change them.

The third part is *Can we get along economically?* The economy has been an unending source of conflict, as, after thousands of years, we still struggle to provide enough for everybody. Today's economic systems, Capitalism and Communism, were the economic technologies of the Industrial

Revolution era. That era is drawing to a close, and those systems are now obsolete. If we play our cards right, AI and personal manufacturing will usher in the next era, which we call *Makerism*.

In *Can government help us get along?* we rethink the role of governments as a vehicle for making collective decisions in society. Representative democracy as practiced in the USA is considered the "gold standard", but like the gold standard for money, maybe there's now another way. We present a proposal which we call *Reasonocracy*, inspired by the collaborative processes of the scientific community instead of the competitive, power-based processes of today's governments.

Finally, in *How can we get along in....?* we give our prescriptions for making a more cooperative society in a number of specific areas. Education. Justice. Guns. Transportation. Infrastructure. These flesh out some concrete solutions to contemporary issues, inspired by our principles.

Backing up our story

References in this book are identified by an author and year in square brackets, e.g. [Fry 2017]. Where an online site does not give a date, the year of our access is used. In online versions of this book, you should be able to click on the reference to take you to the referenced site, article, or book. In print versions, you can look up them up on our online site, http://www.whycantwe.org.

The clock is ticking

The world is not on a sustainable path. Whether current trends end in chaos or not is up to us. The good news is that we think there is, indeed, a sustainable path. Let's take the first steps onto it.

Join us on a journey through solution-space, where we challenge the assumption that adversarial and competitive structures are necessary to get things done.

Part 1
What keeps us from getting along?

Chapter 1
Jailbreaking the Prisoner's Dilemma

I (Lieberman) was 10 years old at the time of the Cuban Missile Crisis, in 1962. It really scared me, because it seemed like there was a real possibility that a nuclear war might bring the world to an end. Russia had set up nuclear missiles in Cuba, 90 miles from Florida. US President Kennedy issued an ultimatum that they be removed. Or else.

Was Kennedy really prepared to start a nuclear war to back up his threat? Would Russian President Khrushchev and Cuban President Castro launch the missiles preemptively, or in retaliation? Tension was high. Many were seriously expecting a nuclear war, because the normal behavior of political and military leaders is: once you make a threat, you have to be willing to back it up.

Most of all, though, everybody feared the fragility of the situation. Nobody thought Kennedy nor Khrushchev had any real intention of starting a nuclear war. But we realized that if they went through the normal political and military processes, one misstep, and nuclear war would be the result.

In school, I remember going through air-raid drills, where we'd be told to crawl under the desks in the event of sirens announcing impending war. But I grew up in New York. Even as a child, I knew enough to realize that if the Russians dropped even a single bomb on the Empire State Building, considering where we were, we'd all be vaporized. The air-raid drills were useless. It was at that moment that I came to the realization: adults didn't know what the heck they were doing.

The discussions on TV about the situation sounded insane. Nobody proposed any sensible way out. The first job of politicians and military leaders was supposed to be to assure that citizens of their countries were safe. But at that moment, it seemed like the leaders themselves were the biggest imaginable threat to the world's citizens. There must have been something wrong with all the political and military processes that were supposed to protect us, if they led us to the absurdity of that point.

What was wrong was that the leaders did not recognize that their interactions formed a particular kind of pattern that was heading to disaster. We're going to tell you about that pattern: *The Prisoner's Dilemma.*

Introducing the *Prisoner's Dilemma*

In this book, we're going to talk about a wide variety of social and economic problems and propose some solutions. But there's a unifying theme—the tradeoff between *cooperation* and *competition*. The thesis of this book is that the root cause of many societal problems is getting this tradeoff wrong.

The political Right extols the virtues of competition; the Left extols the virtues of cooperation. Neither side acknowledges that sometimes it's better to cooperate, and other times, it's better to compete. So admitting that fact is the first step. But once we get that far, there's still the problem of how to determine whether to cooperate or compete in a particular situation.

In this chapter, were going to give you some tools for how to think about the issue. We'll show you an important concept from mathematics, but don't worry if you don't have a strong math background, it's not difficult. We will walk you through it step-by-step, and there are only a couple of very simple formulas.

There are also a number of very good online videos that use animation to teach you the basics. A good one is from *Scientific American* [Moyer 2012], and a simple search on YouTube for "Prisoner's Dilemma" will find several others.

Even mathematicians, though, don't fully appreciate the extent to which this pattern really does describe many real-world situations. They don't always connect the dots between what they learn from the mathematics, and its implications for society. That's our job in this book.

If you're already familiar with the Prisoner's Dilemma, you might be able to skip ahead to the section *We're All Prisoners of Reality*, below, but a refresher certainly won't hurt. Now, let's start.

The reason it's called the Prisoner's Dilemma is because of the following story, used in the literature to present the problem.

> *Two suspects, Bonnie and Clyde, are arrested and held in prison by the police. The police have insufficient evidence for a conviction, and, having separated both prisoners, visit each of them to offer the same deal:*
>
> *If one testifies for the prosecution against the other and the other remains silent, the betrayer goes free and the silent accomplice receives the full 5-year sentence. If both stay silent, the police can sentence both prisoners to only one year in jail for a minor charge, like possession of a weapon. If both betray each other, they both will receive 3-year sentences. Each prisoner must make the choice of whether to betray the other or to remain silent. However, neither prisoner knows for sure what choice the other prisoner will make. So the question this dilemma poses is: What will happen? How will the prisoners act?*

Suppose you're Bonnie. The choice you have is either to stay silent, or to betray your partner, Clyde. What do you do?

Of course, emotionally, your fellow suspect is probably your friend, and you have feelings of loyalty to him. You don't want to play the snitch. The cop is offering you the deal in the hopes that your selfishness will overcome your feelings of loyalty. But the point of this story isn't to depict an emotional struggle between loyalty and selfishness. For the moment, we'll put aside thinking about the emotional impact of the situation.

We'll focus on the question of what might actually be in your self-interest, in the narrow sense of which choice is more likely to result in less jail time for you. Then, we'll look at how the situation shapes up for both you and your partner.

We figure it out using a branch of mathematics called game theory, because it applies both to games like chess and poker, and also to decision-making situations in real life.

Keep in mind that game theory is a way of abstracting the situation by pointing out the mathematical pattern in the story. That way, we can apply the lessons learned to many other situations, whether or not prisoners are involved.

The pattern is solely about whether the sentences the prisoner gets are better or worse, depending upon the choice they make. It's the pattern that counts, not the story itself. The story is only there to motivate, to help you think about it. So you can't get the prisoners out of their dilemma by suggesting, e.g. that they bribe the cops, hire a good lawyer, accuse a third party, have their friend bake a cake with a file in it, tunnel out like El Chapo, etc.

The Prisoner's Dilemma, one step at a time

We'll figure it out, step by step, by thinking about each combination of choices by both prisoners, then comparing them. With a little patience, you'll learn about one of the most remarkable paradoxes in all of mathematics!

Imagine you're Bonnie. The choice you have to make is whether to *stay silent* (cooperate with your partner Clyde, to foil the prosecution); or to *betray* Clyde, in the hopes of getting a lighter sentence. Mathematical game theory uses the term cooperate for choices like staying silent (cooperating with your partner and not cooperating with the cops). It uses the term *defect* for a betrayal, or any kind of choice that is competitive (or uncooperative) with your partner.

To make it easy to follow, we'll draw a set of diagrams, each representing a combination of choices in the shape of a "V". The tops of the two arms of the V represent the choices by each of the prisoners. The point at the bottom of the V represents the outcome, the number of years in jail each prisoner gets. The two arms of the V "cause" the result at the bottom point. We give each particular situation a name, which appears in the middle of the V. The best way to follow the explanation is to look at the picture, read the text, then go back and look at the picture again.

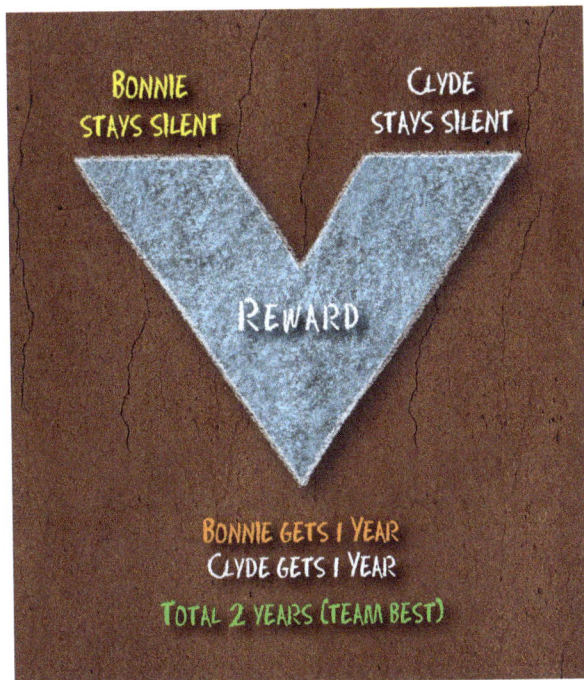

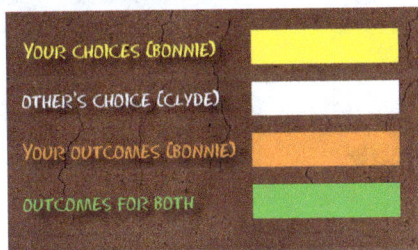

Here's our first example. Suppose both Bonnie and Clyde stay silent, that is, they both refuse to accuse the other one of committing the crime, (while maintaining their own innocence).

In this case, according to the original Prisoner's Dilemma story, they both get one year in jail. If neither confesses, the best the prosecution can do is convict them on a lesser charge, like possession of weapons. This is the best the both of them can do, as we shall see (it's not possible to get both of them off completely). We call this situation the *Reward*.

Now, as Bonnie, you think, "Suppose, instead, I betray Clyde?", accepting the prosecutor's deal. Remember, we're still assuming, for the moment, that Clyde is staying silent. That would lead to the following situation:

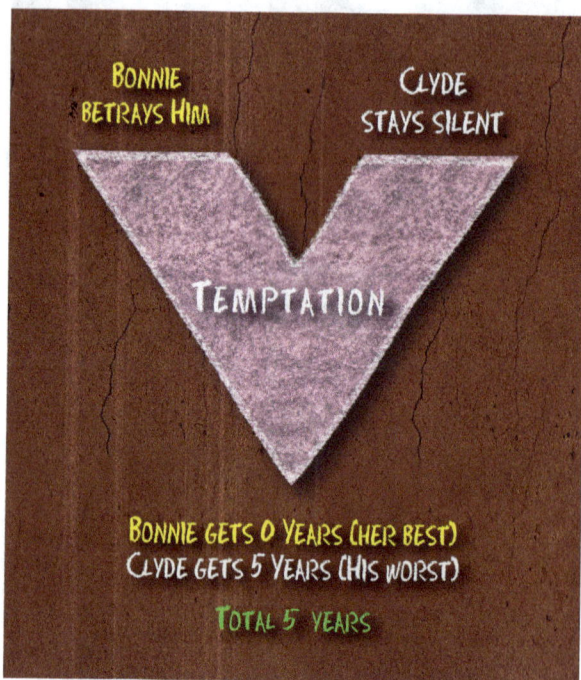

BONNIE BETRAYS HIM   **CLYDE STAYS SILENT**

TEMPTATION

BONNIE GETS 0 YEARS (HER BEST)
CLYDE GETS 5 YEARS (HIS WORST)

TOTAL 5 YEARS

In this case, you get off scot free, but poor Clyde is left holding the 5-year bag. Because you are tempted to betray your partner by the promise of freedom from jail, we'll call this situation the *Temptation*.

Now, let's look at both the Reward and Temptation situations together.

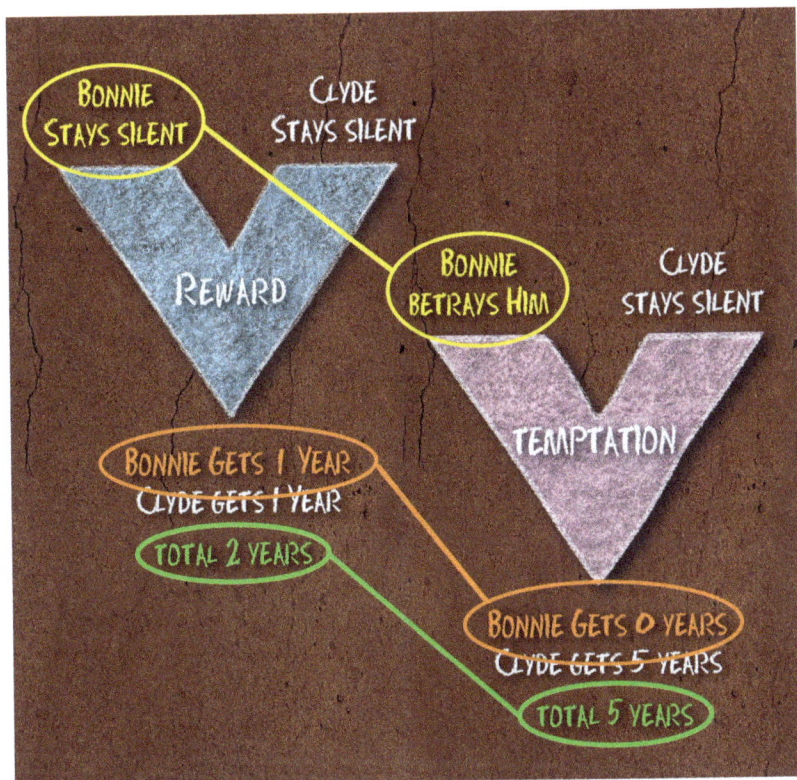

In this diagram, we connect Bonnie's choices (silence or betrayal) in a light yellow (in color versions of this book). A darker orange is used to connect the prison sentence Bonnie gets in each case. In green, we show the "team score" that adds up both Bonnie and Clyde's prison sentences. That way, it's easy to visually compare them across the V's.

Comparing the two situations, which one is better for you? If you stay silent, one year in jail for you, as shown in the upper left V, the Reward. If you betray, you get zero years. That's obviously better than one year in jail. So, (if you can get past your guilt in ratting our your partner), you give in to temptation, and betray him.

But remember, all this is assuming you trust Clyde to stay silent. But suppose, instead, *he* betrays *you*? Then what?

The next diagram introduces two new situations. In both of these, as shown at the upper right of each V, Clyde betrays Bonnie. At the upper left of each V, as before, we have Bonnie's choices.

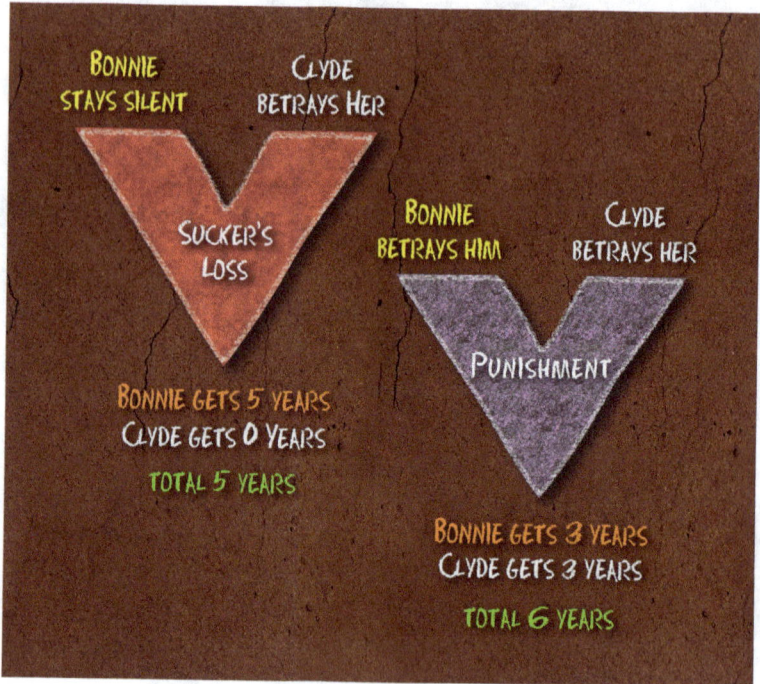

BONNIE
STAYS SILENT

CLYDE
BETRAYS HER

SUCKER'S
LOSS

BONNIE
BETRAYS HIM

CLYDE
BETRAYS HER

PUNISHMENT

BONNIE GETS 5 YEARS
CLYDE GETS 0 YEARS

TOTAL 5 YEARS

BONNIE GETS 3 YEARS
CLYDE GETS 3 YEARS

TOTAL 6 YEARS

You can choose to stay silent. In that case, you will be the *Sucker* who has to serve hard time, the 5 year sentence, while Clyde, the lucky devil, walks. The betrayer serves no jail time.

Or, you can both betray each other, and then you both get the *Punishment* of the intermediate sentence. Three years is worse than 0 or 1, but still not as bad as 5 years. So, again, unfortunately, you decide to betray Clyde.

Let's step back and look at the overall situation, with all four cases displayed.

Your task, as Bonnie, is to choose between the top row and the bottom row. In the top row, you stay silent in both the Reward and Sucker's Loss situations. In the bottom row, you betray Clyde in both the Temptation and the Punishment situations.

Now, we compare prison sentences *vertically* in the grid. On the left side, we're comparing "Bonnie gets 1 year" at the top right square, to "Bonnie gets 0 years" on the bottom right. Fewer years is better, so the bottom row, betrayal, is preferable.

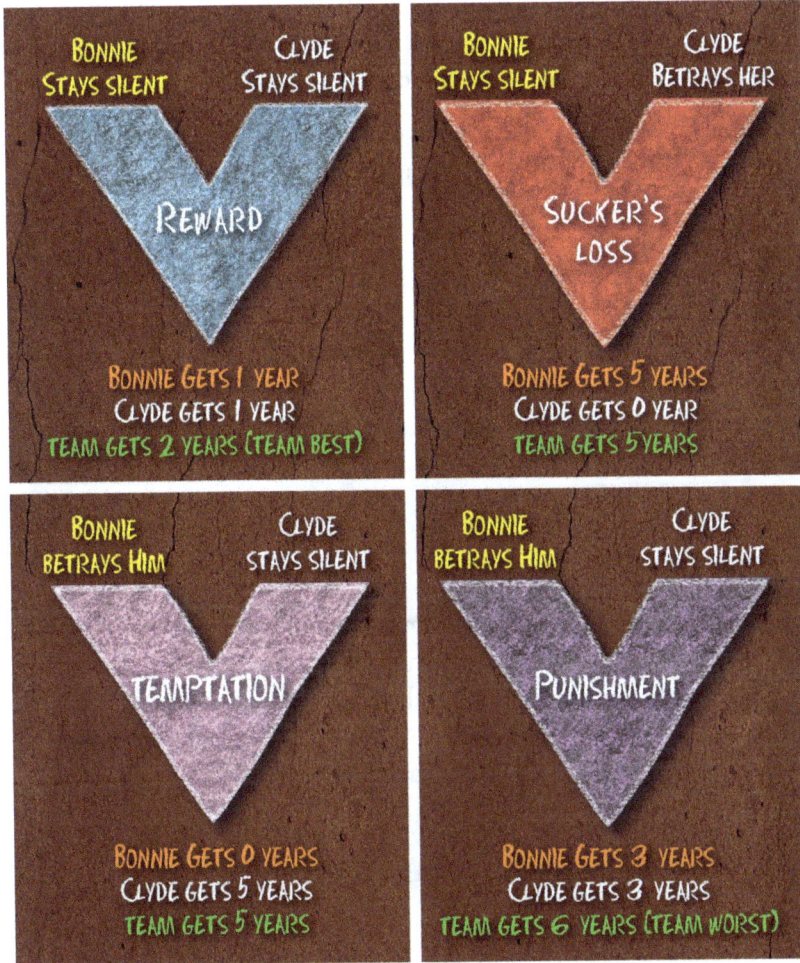

On the right side, we compare "Bonnie gets 5 years" at the top right to "Bonnie gets 3 years" at the bottom right. Again, betrayal wins out. In both cases, the bottom row choice is better than the top row choice. So, no matter what Clyde chooses, if you, Bonnie, act selfishly, you will betray him. Tough luck, Clyde.

But wait a second—isn't Clyde going to go through exactly the same sort of reasoning? Won't Clyde come to the conclusion that *he* should betray *you*, by exactly the same reasoning?

Then, what we're left with, is the Punishment situation. Inevitably you both betray each other, and wind up with the 3 years in jail, apiece. But notice that this outcome is actually the worst situation for the team, with 6 years in the clink, total. The worst outcome!

If only both of you had decided to stay silent, you'd both have gotten off with only a year apiece, the best situation for the both of you. What a shame!

That's the Prisoner's Dilemma. It's a situation where, if you consider your options selfishly, you would choose *not* to cooperate with your partner, because it would seem better for you, regardless of what your partner does. He makes the same choice, and the result is that both of you *defect*. But if you had both cooperated, the situation would have been better, collectively, for both of you!

It may seem paradoxical, but it's something that often happens in the tradeoff between competition and cooperation. What went wrong?

If you just consider the situation from a *local* point of view (your own), you come up with an answer that doesn't achieve the best team score. But if you consider the *global* situation (both your own and that of your partner), you can do the best for everybody. And, as we'll see later, cooperation is your best strategy in the long run. Take one for the team.

What goes wrong, when both of you choose, selfishly, to betray each other? We can summarize this situation as:

> *Think locally, fail globally.*

The obvious cure, (as one of our favorite slogans puts it) is:

> *Think globally, act locally.*

That's where we are with the choice between competition and cooperation in politics, economics, and society. Too many individuals and organizations consider a situation from their own local point of view, and make a bad choice to compete with others they interact with. If they considered the more global situation (how their choices affect themselves and others), and, of course, if they could recognize in their situation the pattern of the Prisoner's Dilemma, they'd choose to cooperate. Things would be better for humanity as a whole. That's the lesson of the Prisoner's Dilemma.

Let's run the numbers

You may notice that the cops had to be really clever about choosing the number of years they offered the prisoners in the plea bargain, in order to trick them into the *Punishment* scenario. If there were different numbers, things would turn out differently.

Prison sentences are a bad thing. To compare the desirability of various situations, we have to keep in mind that *longer* sentences are less desirable. So we'll turn the numbers around, using negative numbers of years to describe the prison sentence. You could think of it as the number of years you *lose* from your life. We'll show the desirability by the length of the bar next to the prison sentence.

If we stack up the prison sentences in the order of their desirability for you, Bonnie, it would look like this:

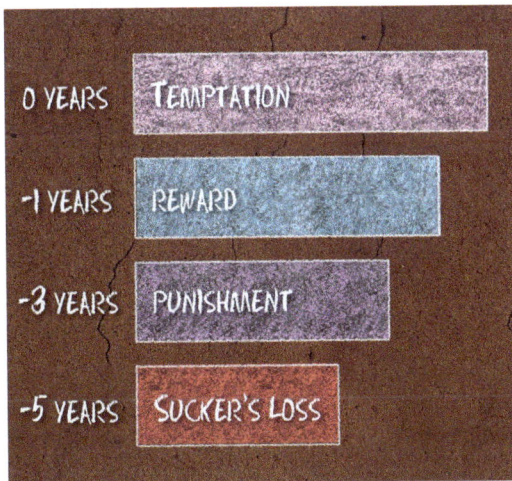

To help you remember the order, we're going to give this formula a mnemonic name: *The TRaPS Inequality*.

The **TRaPS** Inequality:

Temptation > Reward > Punishment > Sucker's Loss

T>R>P>S

The Prisoner's Dilemma "traps" you into defecting when you should be cooperating.

By definition, the term *Prisoner's Dilemma* refers *only* to those situations that satisfy the TRaPS Inequality. Later, we'll see other kinds of situations that can occur.

This is really the only math formula you have to know to understand the Prisoner's Dilemma[2]. We told you the math wouldn't be that hard!

The fact that game theory recommends that each partner defect against (betray) the other in a single Prisoner's Dilemma game is a bummer. It might lead you to a pessimistic view of human nature, that it is inevitable that people will prioritize their own welfare over that of others, and the world would descend into a cesspool of selfishness. Indeed, that really happens in geopolitical situations, as we'll see in our analysis of the Cuban Missile Crisis.

But there's hope for humanity in the story of the Prisoner's Dilemma. In the 1980's, evolutionary biologists found another way of thinking about the problem that leads to a much more optimistic viewpoint. Stick with us.

---

2   Sometimes, though, another mathematical condition is added,
(2 * Reward) > (Temptation + Sucker's Loss)
to avoid a borderline case where players can alternately cooperate and defect. We're just telling you, so knowledgeable readers won't nitpick our contention that the TRaPS Inequali-ty is all you really need to know. Most situations of interest satisfy this condition as well, so we won't have to pay attention to it.

From MADness to cooperation

The Prisoners' Dilemma has a fascinating history. It was first discovered in 1950 by Merrill Flood and Melvin Dresher [Flood 1952], who worked at the Rand Corporation. They were working for the US Defense Department, studying scenarios of the possibility of nuclear war, as this was the height of the Cold War. As we'll see in our chapters on *War*, this scenario does indeed apply to many situations involving potential and actual war.

Though their original intent was to describe a theoretical problem, they soon realized the relevance of their work to the political situation at the time. It showed the folly of the policy of *Mutually Assured Destruction* promoted by the military at the time. The aptly-named MAD was the idea that merely the threat of nuclear war would be sufficiently abhorrent as to prevent nuclear war. Some even went so far as to think that nobody would dare start a conventional war because of the risk they'd provoke a nuclear war.

Flood and Dresher's work pointed out that if things ever managed to satisfy the TRaPS Inequality, we'd risk winding up in a situation where nuclear war and total destruction would become almost inevitable.

Ironically, Flood and Dresher's argument was completely ignored by the Cold War military and politicians for whom they were ostensibly working. On one hand, you could say that MAD "worked" because, fortunately, we managed to (just barely) avoid the nuclear disaster that Flood and Dresher warned about. On the other hand, though, we came pretty damn close. Recent historical investigation has uncovered just how close we really were. Khrushchev's memoir said that Castro had asked him to launch a nuclear attack on the US, a request that he denied because he disdained Castro personally, as "hotheaded" [Pear 1990].

Let's return to the Cuban Missile Crisis, with which we started this chapter, and see how it plays out with a Prisoner's Dilemma analysis. In this situation, either country could cooperate (by removing missiles threatening the other), or defect (by launching (or even threatening) a nuclear strike.

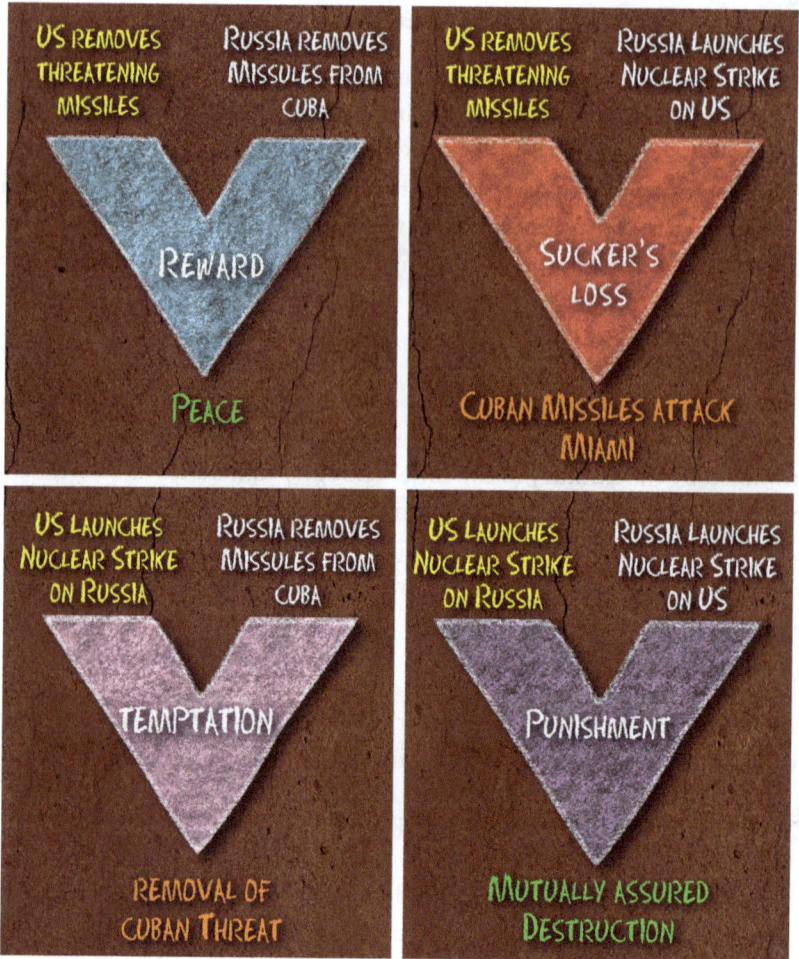

Fear of being a victim of a first-strike nuclear war (*Sucker's Loss*) would seem worse than even the possibility of both sides being destroyed by each other (*Punishment*). The satisfaction of avoiding war (*Reward*) might not be as compelling as the *Temptation* to launch the first strike. A first strike would potentially knock out the retaliatory capability of the other side. The inevitable result, though, would be the *Punishment* of mutual destruction.

Eventually, though, it seems Kennedy made a secret deal that removed missiles in Eastern Europe and elsewhere that were threatening Russia, in exchange for removing the Cuban missiles. On a more black-humorous note, the paradox was later satirized in popular movies, for example, *Dr. Strangelove* and *War Games*.

From war to evolutionary biology

The Prisoner's Dilemma remained a mathematical curiosity, until it was picked up in the 1980s, amazingly, by evolutionary biologists.

The biologists were trying to understand the following puzzle: Evolution is based on competition, as at least a cursory reading of Darwin might indicate, encapsulated by the slogan, "Survival of the fittest". So, how is it possible for cooperation to evolve in human societies?

Until that time, conventional evolutionary theories had a hard time squaring the competitive nature of "survival of the fittest" with the observed tendency of organisms to (at least, sometimes) cooperate. Multi-celled organisms grew out of cooperation between single cell organisms. Mitochondria started out as independent cells, then became the energy source for larger cells. There are many examples in nature of symbiosis between plants and animals. Modern discoveries such as the intestinal microbiome continue to astonish at the complex nature of the cooperation between organisms.

By the 1980s, computer simulations were being used by evolutionary biologists, so Robert Axelrod of the University of Michigan decided to run a tournament of simulated agents, so he could investigate the behavior of the Prisoner's Dilemma in a population of agents.

One novel aspect is that the game is *iterated*. Players could play again and again, with the same partners or different ones, and have a memory of what happened in the past and use it to affect their strategies (though they weren't allow to communicate with each other, to simplify the analysis). This *Iterated Prisoner's Dilemma (IPD)* is the really important case.

Each iteration of the simulation was a round-robin tournament of pairs of agents playing two-player Prisoner's Dilemma with each other. Each agent was controlled by a program that made the decision to cooperate or defect. Some agents had random strategies, but he also invited researchers to submit their own strategies, which would participate in the game. Agents who had successful strategies could "reproduce" within the simulation, so that more agents using that strategy would appear in the next round. Scientists could then see how the population of agents would evolve.

Would the most ruthless, competitive agents win, or would the population of agents learn how to cooperate with one another? This led to Axelrod's 1984 book, *The Evolution of Cooperation* [Axelrod 1984], which formed the basis of the modern understanding of the problem.

The results were startling. The upshot: When the game is repeated, the best strategy changes from defection to cooperation! So, indeed, cooperation can evolve. We're not stuck with the pessimism of game theory telling us to defect in the single-shot case. What a relief!

Why is that true? Well, in the single shot case, you may indeed gain an advantage by defecting. But that advantage comes at a longer-term cost— it will affect how that partner plays with you next time. Strategies that do best in the long term are those who choose to cooperate with a fellow cooperator.

So if you defect, it makes you a less desirable partner, and makes it more likely that you'll be playing with someone who will defect against you. That'll hurt you. This speaks to a short-term vs. long-term tradeoff, and we'll consider many such tradeoffs later the book. As long as you "care enough" about the future, you'll cooperate.

There were many, many variations of strategies tested, and work continues to this day investigating all the possibilities. In Axelrod's initial work, a very simple strategy, submitted by Anatol Rappaport, turned out to be surprisingly successful in a wide variety of situations (but not all): Tit for Tat (TfT). TfT always chose to cooperate on the first round. From then on, it simply mimicked what its partner did on the previous round. Surprisingly, it won over more complex strategies.

What can we learn from the success of Tit for Tat? TfT succeeds, not by defeating its partners (since it imitates its partners, it won't do much better or worse), but by eliciting cooperation from its partners. On one hand, you might think of TfT as the reciprocity rule that most religions have, to encourage followers to have empathy for others.

On the other hand, you might think of it as the equally Biblical "an eye for an eye, a tooth for a tooth" since it "takes revenge" whenever the opponent defects. But the implications of the revenge view were well understood by the game theorist Tevye, a character from the play *Fiddler on the Roof*: "That'll leave the world blind and toothless". Modern interpretations of the Biblical phrase suggest that the Hebrew wording is better read as *"only* an eye for an eye", counseling against *disproportionate* punishment. Modern variants of TfT insert additional forgiveness, even randomly, to avoid endless cycles of defection. This makes the system more robust in the presence of uncertainties.

In a Justice system, the cost of administering justice is often borne by everyone in society, so everyone loses when one person commits a crime. This is unfair, but worth it, considering the inefficiency of lawlessness. Attempts to reduce the cost of Justice, say, by having a dictator decide, often lead to overall worse outcomes for society. Cooperation is not free, it's just better than the alternatives.

More generally, the properties that make TfT and other successful strategies work, are that they are nice (never the first to defect), forgiving (will sometimes return to cooperation after a defection), and provokable (won't always cooperate in the face of continual defection). Understanding principles like this gives us hope that we can develop strategies for Prisoner's Dilemma situations that better lead to cooperation.

Astonishingly, it took until 2013 for somebody to get the idea of actually trying the Prisoner's Dilemma experiment with real prisoners [Khadjavi 2013]. And it turns out that real prisoners are smarter than game theory mathematicians—they choose cooperation, even in one-shot cases, far more often than the mathematics would predict.

No doubt this was because the emotional aspects of the situation (which, you'll recall, we said we were going to put aside when we first told the story) kicked in. Perhaps, also, it was significant that this particular experiment was conducted with only female prisoners. Emotions like loyalty and empathy may well serve an under-appreciated function—to get us out of making poor choices in Prisoner's Dilemma situations!

When competition beats cooperation

Everything also has its flip side. It's important to note that, if the TRaPS Inequality doesn't hold, we don't have the Prisoner's Dilemma. We can even get the opposite situation, where cooperation fails and competition succeeds.

For example, imagine that our cops think that the two prisoners have Mafia connections. They say, "We're going to set up a competition between you two for who can give us the best evidence to convict your Mafia Godfather. If you both stay silent, we'll throw the book at both of you, regardless of your guilt or innocence. The guy who gives us the best evidence that results in convicting the Godfather will go scot-free, and the other guy will take the heat. Whaddya say?"

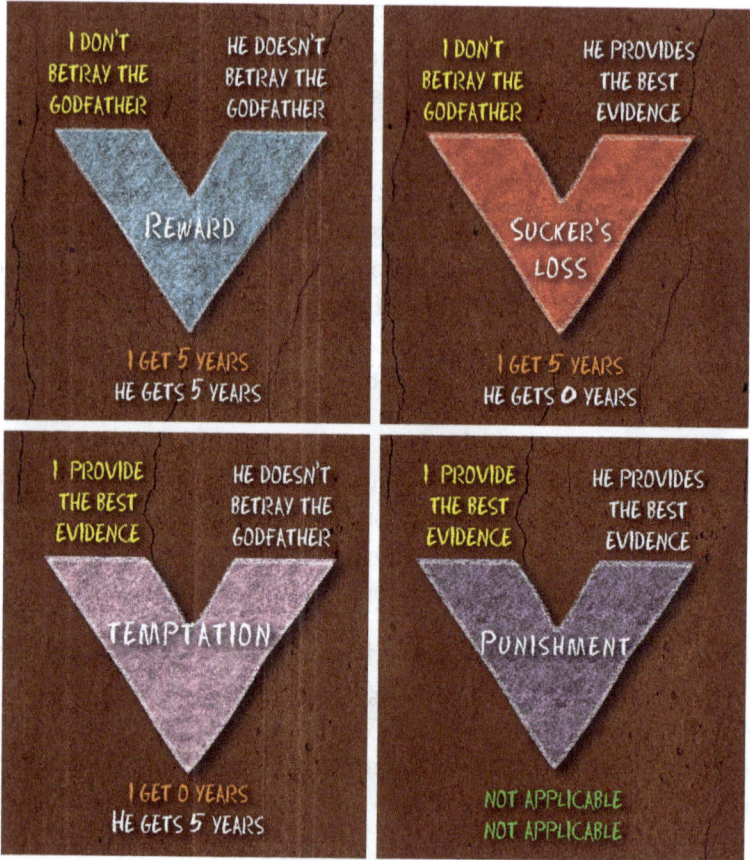

I DON'T BETRAY THE GODFATHER / **HE DOESN'T BETRAY THE GODFATHER** — **REWARD** — I GET 5 YEARS, HE GETS 5 YEARS	**I DON'T BETRAY THE GODFATHER** / **HE PROVIDES THE BEST EVIDENCE** — **SUCKER'S LOSS** — I GET 5 YEARS, HE GETS 0 YEARS
I PROVIDE THE BEST EVIDENCE / **HE DOESN'T BETRAY THE GODFATHER** — **TEMPTATION** — I GET 0 YEARS, HE GETS 5 YEARS	**I PROVIDE THE BEST EVIDENCE** / **HE PROVIDES THE BEST EVIDENCE** — **PUNISHMENT** — NOT APPLICABLE, NOT APPLICABLE

Here, the TRaPS Inequality doesn't hold. We're not allowing ties, so there isn't any *Punishment* for mutual defection. The so-called *Reward* for mutual cooperation results in the worst team situation, both individually, and for the team.

In that case, the best strategy for the prisoner is to deliver the best evidence against the Godfather that he can, even if it implicates his partner. Cooperation (and here we mean cooperation with your partner, not cooperation with the prosecution) loses, competition wins.

Garrett Hardin [Hardin 1968] wrote about the *Tragedy of the Commons*, a classic exposition of the kind of situation where cooperation around a shared resource can fail. He described a situation where many people held a resource in common, for example, a pasture to graze animals. Each animal owner gets the full benefit of grazing his own animal on the common. If the common is big enough to handle all the animals, everything is fine.

But if the common pasture isn't enough to feed all the animals (a situation of *scarcity*), the inevitable result will be that the shared resource will be exhausted. The attempt by the whole group to share a resource cooperatively with one another results in failure, unless—by goodwill, persuasion, or some mechanism that enforces it—everybody grazes only their fair share. We'll return to the issue of scarcity later, where we'll see that scarcity both causes cooperation to fail, and makes cooperation more difficult.

Hardin's situation was essentially a multi-player Prisoner's Dilemma. But he wrote in 1968, before the full impact of the Prisoner's Dilemma was really widely known, and he didn't reference the original work.

When leftists promote the virtues of collective cooperation, Libertarians are quick to counter with the Tragedy of the Commons, to remind them that shared resources don't always work. They recommend a regime of private property, which (at least, in a relatively even wealth distribution) allocates resources amongst participants to avoid exhaustion.

So, if the numbers can go both ways, either favoring cooperation or favoring competition, then which is it? That, of course, depends on the details of the situation. In the upcoming chapters we'll explore some of the variables that affect the outcome (we've already mentioned *scarcity*). But our general argument is that, due to technological change, these variables are moving in the direction of favoring cooperation over competition.

You may have heard the business term, *coopetition*, a mix of cooperation and competition [Brandenburger 1997]. It refers to situations where, even though businesses may compete with each other, they might still be able to find particular issues on which they can cooperate. Google and Facebook may be competitors, but they can cooperate on things like supporting visas for immigrant engineers. We should encourage companies to be more cooperative, resisting their habitual instinct to be competitive in all things. But we really don't like the way this term is normally used in the business literature. Cooperation and competition aren't considered equally the way they are in the Prisoner's Dilemma analysis; cooperation will only be tolerated in the service of larger, competitive goals. So we won't use this term.

Another difficulty you may notice in applying these lessons to the real world, is that in many cases, we're talking about stuff that's difficult to quantify. If we're talking about people's happiness, or loss due to death in a war, etc. these are intangible quantities and can't be as easily or accurately measured as years in jail.

One response is to try to develop measures that stand in as a proxy for the quantity of interest. The government of Bhutan, noting that the traditional economic measure of the Gross Domestic Product (GDP) ignores many factors, committed itself to trying to maximize Gross National Happiness, by aggregating a variety of quality-of-life measures [Bhutan 2016]. In the case of the Prisoner's Dilemma, it's important to note that the central formula is an inequality, so it is relatively insensitive to the exact numbers for each of the factors. It's their relationship that counts.

We're all prisoners of reality

Let's look at how some of this plays out in some real world situations.

War

One of the most obvious, the most destructive, and the most absurd behaviors, is war. Why do nations go to war? Nobody will admit that they want war. So how can war possibly happen?

If both sides cooperate with each other by eschewing war, both will benefit by having more resources to meet their citizens' needs. Neither really "wants" a war that will result in death, injury, and waste of material resources on both sides.

But they become more afraid of the possibility of being defeated militarily by an unscathed conquering foe, than they are of the situation where they get attacked, but they can attack back. Even if it causes equivalent loss for their own side. Of course, warmongers don't ever think about the loss of human life and cost to the other side.

The official story is usually that the threat of retaliation will deter attackers. But the overwhelming historical evidence is: even when there's a credible threat of retaliation, wars occur anyway. Overconfident adversaries simply dismiss the possibility that they might be defeated. So deterrence rarely works.

To see this, simply ask the warmonger: If there were a significant chance of being defeated, would that deter you from proposing war? First, you'd have to get past their conviction that defeat was impossible. Then they'd say that, even if defeat were possible, it would preserve their honor to "go down fighting".

It's even worse than that—deterrence *can't possibly* work, because military people take an oath that they will persist in trying to destroy the enemy even in the face of their own destruction. That is, both sides have agreed in advance that deterrence is out of the question. If deterrence doesn't work for them, how could they expect it would deter the other side?

Often, just the *possibility* of war engenders such fear in people, that they believe they have to build up an army and weapons in "defense". The military (and its suppliers and supporters) constitute a powerful force with an incentive for exaggerating threats, so that they'll feel needed and valued by the society, and they can profit from military preparations. If that occurs on both sides, then voilà, you've got war.

We've already alluded to the reasoning in our discussion of Flood and Dresher's original work on nuclear war. But let's break it down here and think about the general situation. Each side always has a choice about whether or not to attack (or to take revenge for an attack, which here, amounts to the same thing). Suppose the *Temptation* to achieve military victory (or removing the threat) is perceived to exceed the *Reward* of a peaceful and prosperous society; and the prospect of the *Punishment* of both sides being damaged by a war, bad as it is, is at least considered better than the *Sucker's Loss* of being attacked while not having a military response.

Then we've got Temptation > Reward > Punishment > Sucker's Loss. We've got the Prisoner's Dilemma. It doesn't matter "who started it" (the favored excuse of the playground bully). Off to war we go!

Let's review. (See diagram on page 33.) If both sides cooperate by refraining from war, the result is peace and prosperity. If they both defect (attack), war breaks out and destruction happens on both sides. But neither side knows what the other will decide, so we have to investigate both possibilities. Since it's always possible the other side will attack (or the other side has already attacked), military action may seem more desirable than the status quo, since, if it is victorious, it will deal with the (perceived or actual) threat. A war is bad for both sides, of course, but it might still seem better than being unilaterally defeated by the opponent. At least we fought back, they'll say.

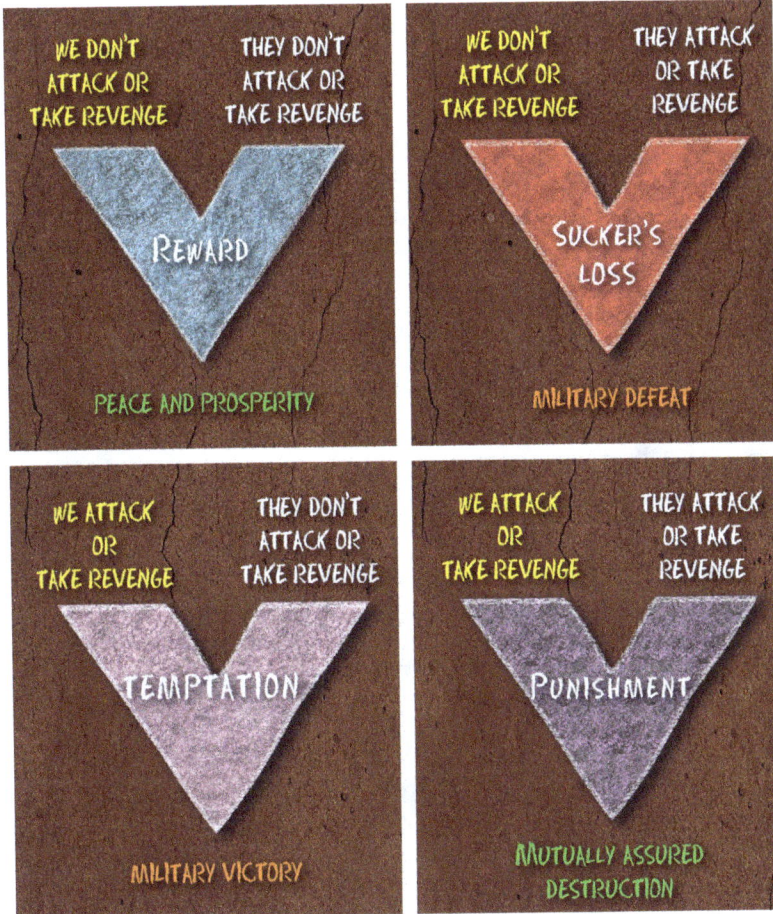

So, for the nation that has to choose either the top row (don't attack) or the bottom row (attack), the bottom row seems to be better in both cases. So it chooses to attack. So does the other side, which is faced with exactly the same choice. Now you know why we have war: Fear of war causes war. We can't emphasize this enough, so we'll say it again, in bold italics.

Fear of war causes war.

Franklin Roosevelt said, "The only thing we have to fear is fear itself". Because the Prisoner's Dilemma hadn't yet been discovered, he couldn't have known how right he was. And if the public *had* understood how right he was, we probably could have avoided all the wars that followed World War II.

Pretend wars and real wars

It's not always obvious, in a real situation, who the prisoners actually are.

Another way to view the situation of war is that there's a *pretend war* between the two nations that are ostensibly fighting each other. But the *real war* is between the military-industrial complex and the citizens, on both sides. So there are actually two real wars. US President Eisenhower warned of this in the speech where he coined the term *military-industrial complex*.

This too, amounts to a Prisoner's Dilemma. The *Temptation*, for a military, is to enhance its prestige with its citizens, to recruit, to increase the military budget, increase opportunities for heroism, etc. It can do this by exaggerating the potential threat, thwarting attempts to negotiate treaties with the other side, making demands for resources, stirring up fear. Military contractors make large contributions to political candidates likely to support war, especially in districts where these contractors hold economic power.

Military people feel most useful and engaged when actually fighting threats, what they signed up to do. This gives them a cognitive bias towards considering threats more likely than the average person. Military commanders stir up recruits' anger and aggression, to make it more likely they'll take action instinctively should the need arise. This becomes a self-fulfilling prophecy.

The *Sucker's Loss* for the military-industrial complex is to be considered useless by the population, and have its budget and personnel slashed. That would, of course, be better for the citizens, leaving more resources and reducing death and injury. So the military is actually *defecting* against the citizens. It implicitly *cooperates* with the so-called "enemy's" military, whose citizens are faced with exactly the same choices.

The story told to Americans to justify the Cold War was that Communism was aggressively hell-bent on taking over the world, and it had to be stopped before it was too late: *The Domino Theory*. America was helping out third-world battleground nations by spreading democracy and free markets.

When the Cold War finally ended, the West could finally communicate openly with people in the former Communist bloc. We could ask them, what were they thinking at the time? They weren't discussing strategy for taking over the world. They feared that *American Capitalism* was aggressively taking over the world, and had to be stopped before it was too late [Vorobiev 2016]. The very same Domino Theory. You can imagine how they might get that impression: a Soviet official visiting Angola and seeing people there drinking Coca-Cola, a sight he hadn't seen the previous year.

Even if they weren't trying to take over the world themselves, of course they believed that Communism was a superior system to Capitalism. Russia was just "helping" innocent third-world nations by promoting Communism. Russia had to defend itself against America, the only country in the world who had used nuclear weapons, a proof of the insanity of its leaders. It was the Russians' (paranoid or not) fear of America that led to behavior easily perceived by the American government as aggression.

So much for war.

Capitalism

Now, let's look at commercial competition in our economy. Actually, many business people use the language of war to describe business competition, so the concepts aren't actually all that different.

We live in a Capitalist society, and one of the founding principles of capitalism is competition between companies and competition between products. "Competition brings out the best", capitalists say, supposedly to consumers' benefit.

But that doesn't consider that you might lose out on the opportunity that would have happened had you cooperated, both with the customers and other companies. Neither it is considering the cost of duplicating effort also performed by the other company. So what we've got here is a cooperation/competition tradeoff. What better way to analyze such a tradeoff than the Prisoner's Dilemma?

Commercial competition: Value proposition

Companies are always faced with a choice. They can provide products that provide good value for their customers, charge fair prices and treat their customers well. Or they can provide products that are poor value, overpriced, and fail at customer service. The latter is often more profitable.

Business people say they want to provide fair value and treat customers well, as captured by the business slogan, "The customer comes first". But the *Temptation* for higher short-term profits, driven relentlessly by Wall Street, can sometimes overwhelm their sense of fairness. This is coupled with the fear of the *Sucker's Loss* of being defeated by competitors, or the stock price tanking as a result of low sales.

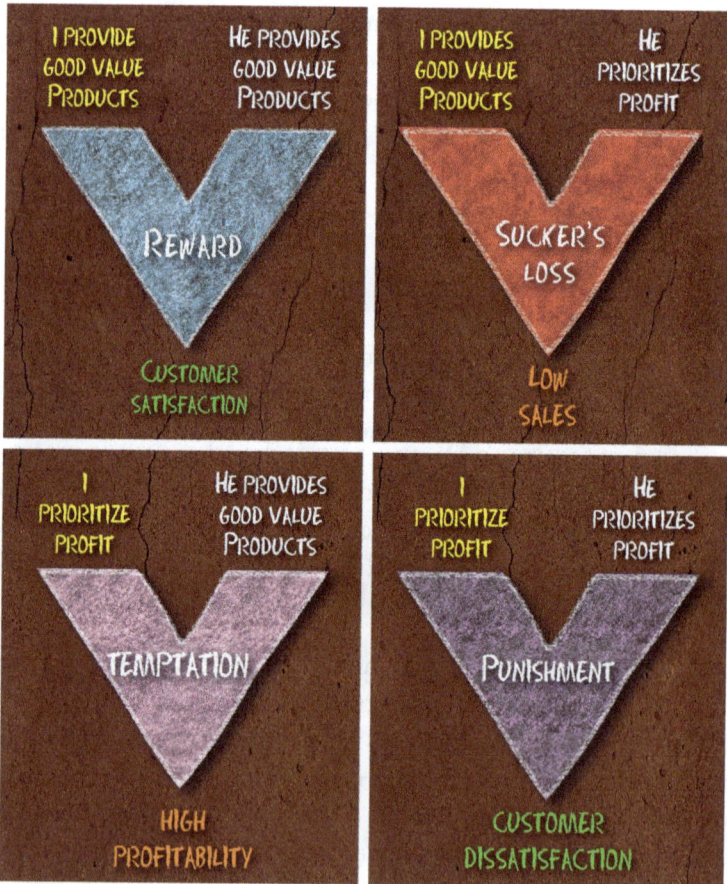

Fear of other companies and Wall St. is the cause of poor value products.

Commercial competition: Advertising

Let's take the example of advertising. Some of the purpose of advertising is to make customers familiar with a product, or to grow the demand for that kind of product, not just the vendor's brand. But in the case of established companies in mature markets, like Coke and Pepsi, consumers already know the brands, and overall cola demand is unlikely to change much. So it's all about market share.

To give you some perspective, the budget of the US National Science Foundation is roughly $7 billion. The advertising budgets of Coke and Pepsi combine to around the same $7 billion. Which does more good for the world?

And, year in and year out, the market shares of Coke and Pepsi don't change very much, so their "return on investment" in advertising is essentially zero. So how does an ad agency convince Coca-Cola to spend on advertising?

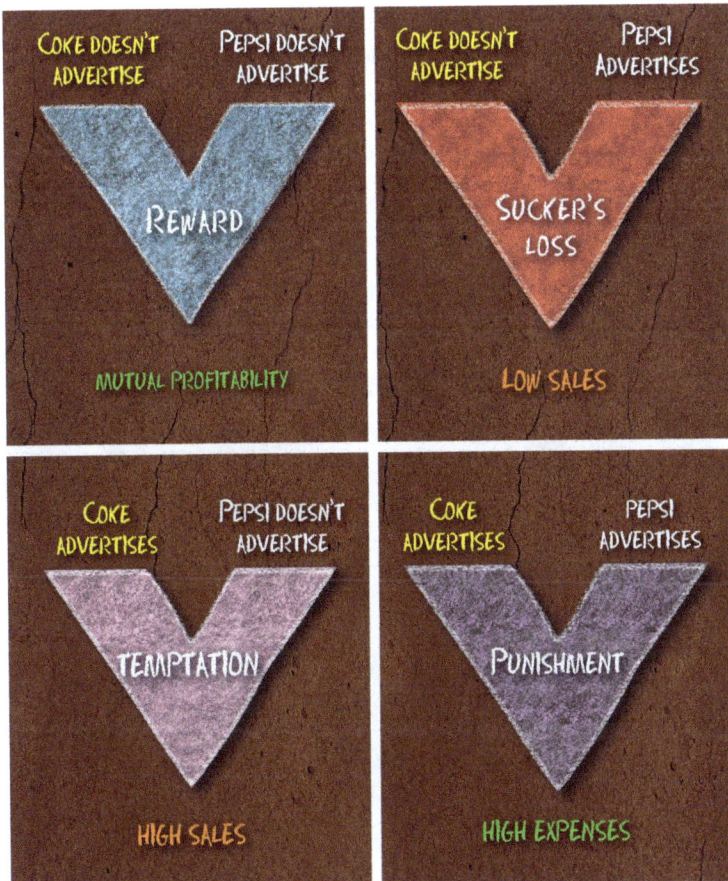

If both companies don't advertise (cooperate), they each have more money. But each company is Tempted by the prospect of success of their ad campaign, and fearful of the Sucker's Loss of the success of the other's company's ad campaign, so they are each stuck with their bloated ad budgets, and the rest of us have to waste our time watching the ridiculous, annoying, lying ads.

Why do we have deceptive and annoying advertising that doesn't do any good for either vendors or consumers? Why do we have expensive sugar water promoting diabetes, almost identically from both Coke and Pepsi? Why did $7 billion go down the drain?

> *Fear of competing companies is the cause of useless advertising.*

Again, we can take another point of view on the situation. The *pretend war* is between competing companies like Coke and Pepsi. But the *real war* is between both companies on one side, and their consumers on the other side. Another *real war* that takes place is between Coke and Pepsi on one side, and on the other, smaller competitors: beverage companies with less market power, startups, supermarket brands, etc.

Environmental pollution

Climate change, pollution, and other environmental crises can also be explained with the concept of the Prisoner's Dilemma. A polluter, like an oil or gas company, gets profit from the pollution, but makes an incremental contribution to global warming. If there's too much of that, ecological collapse is the result.

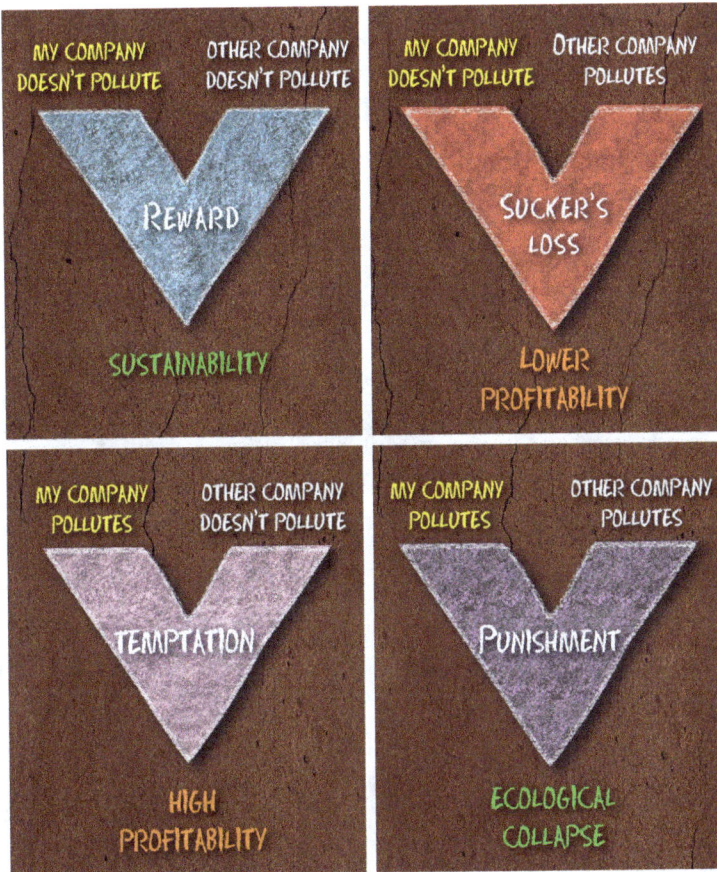

Why do we have unending pollution and no Carbon Tax, which a consensus of scientists and economists agree is a rational response to climate change?

Fear of lower profits is the cause of pollution.

Political gridlock

Let's explain the gridlock in today's US Congress. If both parties cooperated to solve national problems, it would be better for the public, whom they are ostensibly supposed to serve. But the competitive election system, and the organized bribery system, called "lobbying", put such fear in the hearts of elected officials that we get

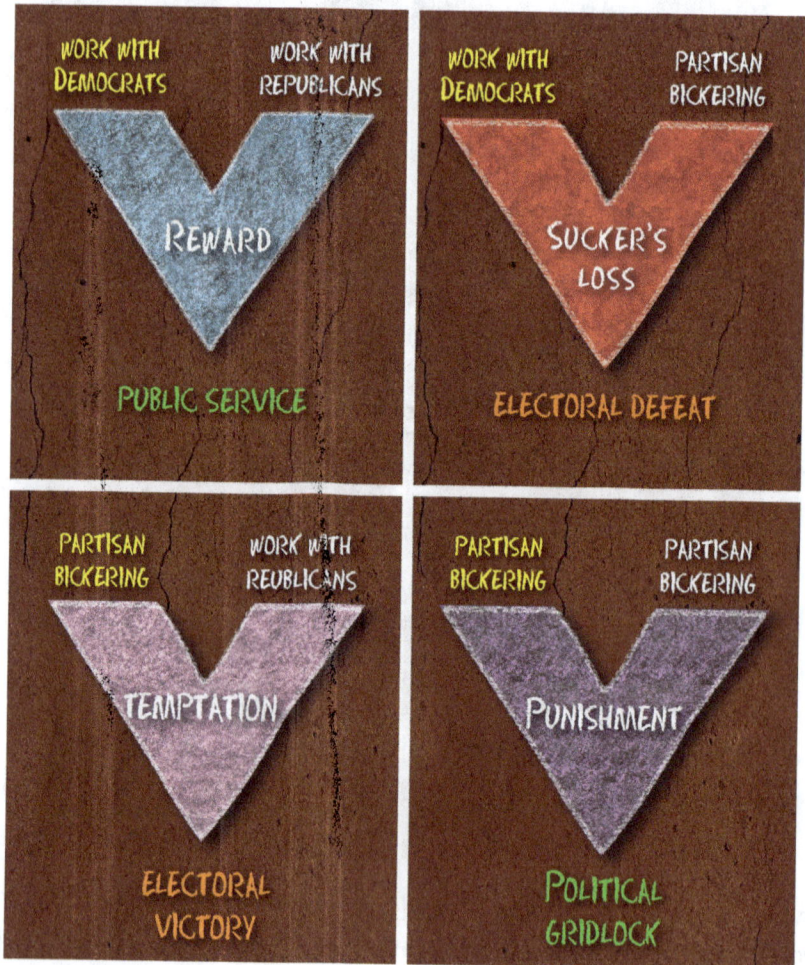

Why political gridlock?

Fear of losing elections causes political gridlock.

Racism, sexism and other forms of discrimination

Even phenomena like racism can be modeled with the Prisoner's Dilemma. [Axelrod & Hammond 03] reported a simulation where a majority group cooperated with each other and defected with the minority group. It has nothing to do with the actual characteristics of the group, only that it is apparent which group a given agent belongs to. It formed a stable pattern. But of course, it led to suboptimal results for everybody.

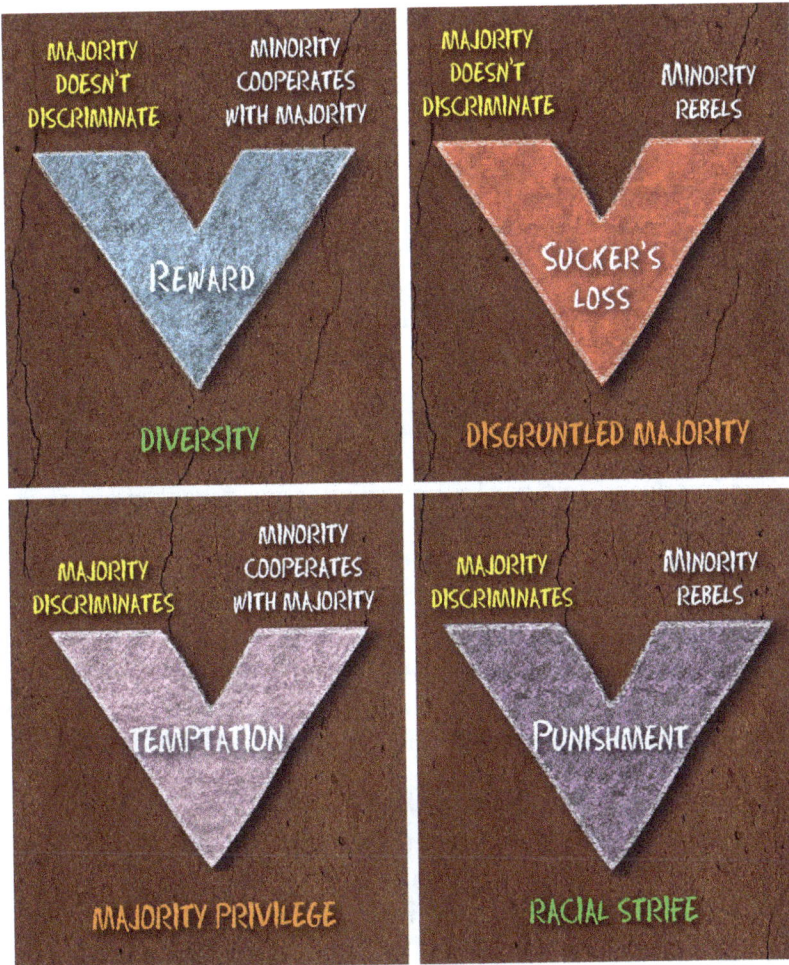

The *Temptation* of establishing privilege for a majority group is that that group thinks it might get some advantage from oppressing the minority. It's fearful of being taken for *Suckers* by the minority, a fear that is usually baseless. Those who advocate the *Temptation* need that fear in order to convince the majority that the discrimination is necessary. But they'll miss out on the *Reward* of a diverse and harmonious society, which is better for productivity and certainly ethically preferable. So we wind up with the *Punishment* of perpetual strife between the factions.

Similarly, any kind of nationalism, tribalism, sexism, homophobia, religious discrimination, etc. etc. is a loss for everybody. So why does it take place?

> *Fear of losing out to a minority causes discrimination.*

There's also another pretend war going on here, as you might have guessed. The leaders of the majority use fear of the minority to gain power over the majority citizens.

Reducing recidivism in the Prisoners' Dilemma

In describing these very diverse problems in similar terms, we hope you can start to see the pattern. And if the problems all have the same pattern, the solutions to these diverse problems also have a pattern.

Basically, in each case, we have to try to make sure that greed for the *Temptation* doesn't overwhelm the hope for the *Reward*; and the fear of the *Sucker's Loss* doesn't overwhelm motivation to avoid the *Punishment*. Then we can make the choice to *cooperate*.

In war, we have to realize that the situation of peace and prosperity, the *Reward*, is the best thing that could possibly happen. We have to resist the *Temptation* of military victory promoted by the warmongers. Our security is better served by *cooperation* with the other side, promoting common interests to achieve peace, rather than fear of the *Sucker's Loss* of defeat.

In commerce, the *Temptation* to produce shoddier products or raise prices might gain them some short-term profitability advantage or favor with Wall Street, but in the long run causes them to miss out on the *Reward* of customer satisfaction, well-being and loyalty. Fighting with other companies for market share at no benefit to the consumer is similarly pointless.

Environmental crises will be just as bad for the polluter as they will be for everyone else. Companies should actually *welcome* regulation, since if everybody is forced to curb pollution, it can remove the fear of losing competitive advantage. Car companies' profits didn't decline after they were forced to introduce seatbelts (and, as a bonus, they had more live customers!).

Politicians should realize that cooperating across the aisle would be their best bet of being able to achieve concrete results they could brag about to their constituency. The *Temptation* of negative ads is that they may cause temporary declines in polling numbers for their opponent. And nobody's fooled, as public approval ratings for Congress descend into the single digits. (Or perhaps everybody's fooled, since re-election rates are high.)

Cooperation across the aisle would better achieve the *Reward* of meaningful public service. And, as civil rights movements expand and succeed, people are gradually realizing that the benefits of racial, ethnic, and gender diversity far outweigh whatever benefits a dominant group feels it might get from maintaining discrimination.

We won't arrive at a single strategy, since it is clear that Prisoner's Dilemma simulations are extraordinarily sensitive to the population of agents, their history, and communication. But we are likely to get increasingly valuable insight into what makes cooperation and/or competition work, and under what circumstances. Understanding Prisoner's Dilemma situations is humanity's get-out-of-jail-free card.

Recap

This has been a lot, so let's review the main points.

- The tradeoff between cooperation and competition can be modeled by a table that shows the results of each possible choice for each participant.

- In some situations, if each participant considers only what's best for themselves, they'll choose not to cooperate, or to compete (called defecting).

- But if you look at the overall situation, it would be better if everybody decided to cooperate.

- This paradoxical situation is called the Prisoner's Dilemma.

- The Prisoner's Dilemma only happens when the numbers in the table satisfy a certain mathematical formula, the TRaPS Inequality.

  Temptation > Reward > Punishment > Sucker's Loss

- This often happens in real life. It can explain situations of
  - War
  - Failures of economic systems
  - Political gridlock and dysfunction
  - Racial, ethnic, and other kinds of discrimination

- We haven't yet found a general solution to the Prisoner's Dilemma, but we know several factors (like scarcity) that make it more or less likely.

- The key to solving many of society's problems is recognizing when this pattern appears, and trying to influence the factors that give rise to it.

Chapter 2
Learn about the Ultimatum Game—
or else

We hope you've gotten some insights from our discussion of the Prisoner's Dilemma. We've got one more game theory concept to talk about, and, don't worry, this one doesn't involve any criminals. But we're not finished with the topic of greed.

It's called *The Ultimatum Game* [Guth 1982]. Like the Prisoner's Dilemma, it starts with a story. There are three participants: The *Benefactor*, the *Donor* and the *Recipient*. You play the Recipient.

The Benefactor offers to give the Donor $100 and says to him:

"Your job is to decide how to divide up the $100 between you two, the Donor and the Recipient. You, the Donor, make the decision. You can divide it up any way you like, but you have to make only one offer. If that offer isn't accepted by the Recipient, neither of you gets anything."

*Now, here's the question: What's the **minimum** offer you, the Recipient, should accept?*

Same ground rules here as in the Prisoner's Dilemma—just one round (we can think about the iterated case separately), no discussion between the Donor and Recipient, and putting emotions aside for the moment.

Most people, if placed in the position of the Donor, would make the simple choice to divide it 50-50 and offer you $50. Most people would consider a 50-50 offer fair, and accept. So what's the problem?

Here's the game theory answer: Any non-zero offer should be accepted by the Recipient. Even if it's only one cent.

What?! That wouldn't be fair!!!

Remember, from the perspective of game theory, we're putting aside for the moment the emotional aspects of loyalty, trust, empathy and fairness that come into play when real people are involved in real situations. We're just considering whether the financial outcomes are advantageous or not. Indeed, when experiments are tried with real people, their sense of fairness does indeed kick in for most people, and the actual offers are mostly close to fair [Christian 2016].

Think about it this way. If the Donor is just a little greedier, they might think, "Since I've got my hands on the $100 right now, I can take a $10 'commission' and the Recipient wouldn't have much alternative but to accept. So I'll offer just $40.". This might still feel acceptable, although the Recipient would probably feel a little annoyed at being taken advantage of.

Now suppose the Donor ups the commission to $20? Still, the Recipient would have to accept. And so on. Even the offer of a single cent would have to be accepted, since the alternative is nothing. Even a cent is better than nothing. That's the Ultimatum.

It's like the Prisoner's Dilemma in that each player can choose to Cooperate (make a fair offer, accept an offer) or Defect (make an unfair offer, refuse the offer). But it's different than the Prisoner's Dilemma in that the game is not symmetric. The Donor has the upper hand.

Here are the choices:

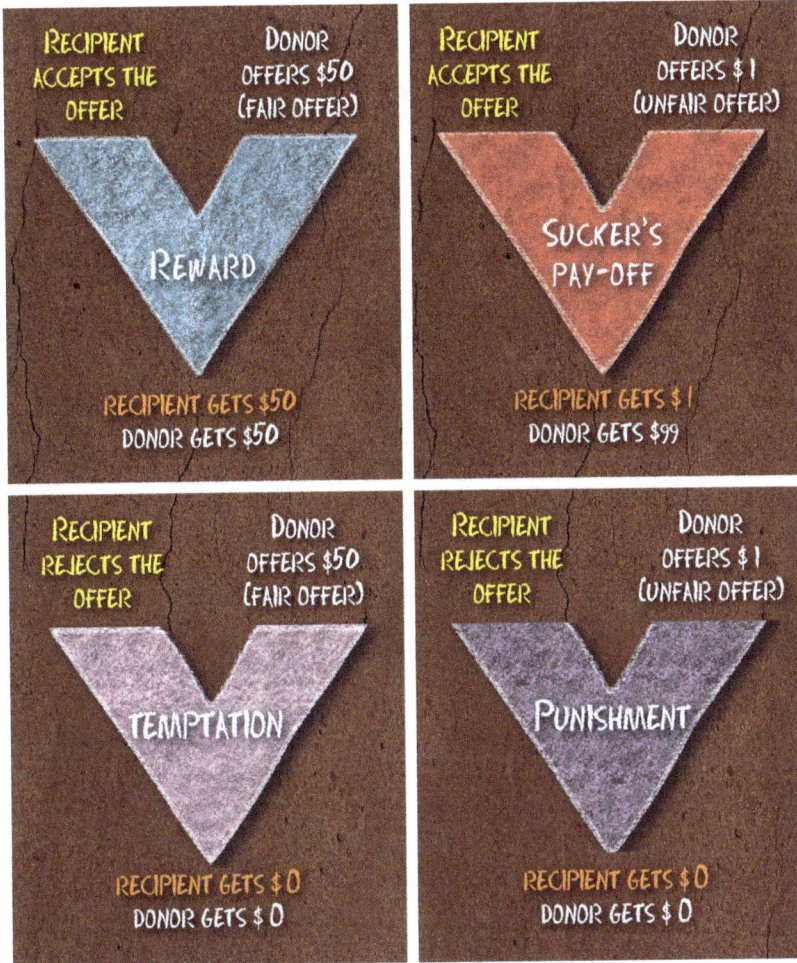

The decision the Recipient has is between the top row (accept offer) and the bottom row (reject offer). In the table, since $50 > $0 and $1 > $0, the Recipient should accept any positive offer, no matter what it is.

But this situation is not the same as the Prisoner's Dilemma. The TRaPS Inequality is not satisfied. Remember the TRaPS Inequality is:

Temptation > Reward > Punishment > Sucker's payoff

But here we've got:

$50	REWARD
$1	SUCKER'S LOSS
$0	TEMPTATION
$0	PUNISHMENT

You might think, "Hey, the Recipient has some leverage, too. They can always decline the offer, in which case the Donor loses out, too. The Recipient could use the threat of declining the offer to negotiate the commission down to a minimal level."

And, indeed, in some real world situations, this happens. When emotions come into play, many Recipients will decline an offer even when it's to their financial advantage, out of spite against being exploited, or out of a sense of standing up for fairness in that situation.

But, like in all game theory scenarios, when we're analyzing the situation, we try to put aside our emotions and speculation about "what the other guy is thinking". Suppose you're the Recipient, but the Donor isn't another person, but a computer program. In that case, you'll realize you can't "teach it a lesson" by refusing, so your decision boils down to either taking the $1 or getting nothing.

Let's look at it from the Donor's point of view. The Donor has to choose between the two columns in the table, the Fair offer and the Unfair offer. Since $99 > $50, the Donor will prefer the unfair offer. The Donor doesn't know how insistent on a Fair offer the Recipient will be. But there's no difference in the case of the Recipient rejecting the offer, where everybody gets zero. So the Donor "might as well try" for the Unfair offer. The Recipient's possible leverage vanishes. Unless the Recipient actually has a viable alternative to accepting the Donor's offer, the Recipient is stuck.

Trickle down theories

Back in the real world, the Ultimatum Game shows us what's wrong with things like "trickle down" theories of economic development. The political Right often advocates policies that directly benefit big businesses and rich individuals. As for smaller businesses and lower-income individuals, the Right claims that they will indirectly benefit through "making the pie bigger", even if they're only entitled to a much smaller share of it. Some grain of truth to that.

But if it's the rich factions who get to decide the split, then what we've got is an Ultimatum Game, where the 1% play the Donor and 99% the Recipient. By the results of the Ultimatum Game, the lower income segments have no choice but to accept anything offered by the rich, no matter how small.

You'll notice that the answer to the question of what's the minimum the Recipient should accept is still, "any positive offer" regardless of the amount the Donor was initially given. Still, anything is better than nothing. Increasing the initial amount from $100 to $200 won't affect what the Recipient gets. Nothing assures that the Recipient's share will be computed as a percentage of the initial outlay. So much for "making the pie bigger".

That's why, for example, we have Minimum Wage laws in the US. How are peoples' salaries set? To quote an ad from a popular negotiation course, "You don't get what you are worth—you get what you can negotiate." In negotiation theory, your leverage in a market is determined by your BATNA—Best Alternative to a Negotiated Agreement [Fisher 1981].

By definition, the people at the bottom of the pyramid have nothing below them, so they have no alternative. No leverage. Marx understood this, a century before game theory, and wrote that the lowest-paid workers' salaries would fall to subsistence level. And once the lowest level is fixed at near zero, then the people just one level up only have the alternative of the lowest level, so they won't fare much better, and so on. Minimum Wage laws prevent this situation, and that's why there's so much pressure now to update the levels to present economic conditions.

The Right does have a counter argument, which is that market forces will always generate another alternative. If workers are unhappy with their wages, some entrepreneur will always arise to provide an alternative. We'll call this counter-argument, the *Imaginary Friend* argument. Markets do indeed provide some incentive for this to happen, and sometimes it does. But how realistic is the assumption that that alternative will always be there when needed?

It's like the old joke among economists:

> *Two economists are walking down the street and one sees a $10 bill on the sidewalk. He reaches down to pick it up, and the other grabs his hand to stop him, saying "Wait! If that were a real $10 bill, somebody else would have already had the incentive to pick it up".*

Incentive, by itself, isn't enough to cause something to happen. The fact that the Recipient is getting a raw deal in the Ultimatum game might indeed provide some incentive for an external entrepreneur to provide another alternative. But if that alternative doesn't actually materialize, the Recipient is still stuck.

How long will it take for the alternative to appear? How much actual, quantifiable effect does it have on the situation? The entrepreneurial mechanism of generation of alternatives is generally far less reliable and effective than the dynamics of the Ultimatum Game. And the rich make money on the spread.

The Right is fond of the image of the lone productive entrepreneur or small business owner. But in the real world, it usually takes a group of people working together to produce significant economic value. So when that economic value actually appears, and the group has to share it amongst themselves, how do we decide who gets what?

In small groups of homogeneous, friendly people, splitting it equally usually satisfies everyone. In larger, more varied groups, however, the power relationships of the Ultimatum Game kick in, so the answer is: The first ones to get paid usually keep most of it, and they distribute the dregs down the line. That's why your Estate Grown Organic Tanzanian Peaberry Coffee costs $18.99 a pound and the poor farmer in Tanzania gets $0.17 per pound.

Imagine that making a product takes a sequential "supply chain" of 10 people, each of whom takes input from the previous worker, and spends an hour of work producing the output to be fed to the next worker. You don't get the finished product until all 10 workers have done their jobs. Each worker's contribution is essential and there are no practical alternatives that could replace someone in the chain. Everybody's contribution is basically indistinguishable.

But now, the consumer buys the product and pays the last worker in the chain, who's the only one that has the finished product. The last worker has all the money and the discretion to divide it up with the rest of the team. That last worker can play an Ultimatum Game with everybody else and offer only the minimum necessary to keep everybody else from rejecting the offer. Enjoy your seventeen cents, Mr. African Coffee Farmer.

Hildalgo refers to this as Topocracy [Hidalgo 2015], where what you receive is determined more by your position in the supply chain than by your effort or the value you produce. The longer the supply chain, the worse it is, which is why *disintermediation*, cutting out the middlemen, is one of the most effective routes for changing the situation.

The Ultimatum Game mixes poorly with another feature of our capitalist economy: the Law of Supply and Demand. The beauty of this Law in an idealized capitalist economy is that it performs *price discovery*—as the economic textbooks say. It sets the price high enough to incentivize the vendor to produce the product, but low enough so that the customer still agrees to buy the product. But that only really works well if supply and demand are approximately in balance. At the extremes of supply and demand imbalance, you have an Ultimatum Game situation. Then, it's Ultimate Game that really sets the price, and the Law of Supply and Demand takes a back seat.

Take for example, the idea of *co-pays* for medical services covered by medical insurance. In the old days, the idea was that insurance, in order to give you peace of mind, would cover the entirety of expenses. Then the idea of a co-pay was introduced, ostensibly to make sure the patient had some "skin in the game" and wouldn't use medical services unnecessarily.

But at what level should that co-pay be? Unfortunately, medical services aren't really like a consumer item like clothing. Generally, when you need them, you need them. You usually only have a very limited discretionary ability to "forgo the luxury", or shop around for alternatives. Attempts to turn medicine into a market by imitating the marketing techniques for other consumer goods are laughable. Let's see, should I choose the medical plan that has the ad with a kindly old doctor smiling at a baby, or the one that has a beautiful young woman running through a field of flowers?

Economists' term for this is that there's little *elasticity of demand*, and therefore, no really effective competition. The choice you're really faced with is "your money or your life".

The temptation for medical service providers is to slowly raise the co-pay to the point "the market will bear"—which is to say, the same level that would be bearable by the customers in the total absence of insurance. The insurance becomes worthless. This hasn't taken place widely yet, but you can certainly expect co-pays to slowly creep up unless government regulators, or industry self-policing groups are vigilant.

That's a general problem with Capitalism. If sellers charge "what the market will bear", then, by definition, prices will always be just short of unbearable. Do we really want to live in a world where everything is always on the edge of being unbearable?

The lesson is, for trickle down theories to work, someone's got to be constantly checking the water pressure.

Chapter 3
Survival of the most cooperative

We're usually not in the habit of agreeing with Creationists. But in their attempts to discredit evolution, they put forward one skeptical argument where we think they might actually have a point. It's about the Darwinian notion of "survival of the fittest".

Who are the "fittest"? Why, they are, almost by definition, those who survive.

If you take survival of the fittest as a hypothesis, then it's hard to make the case that, if a particular trait seems to be manifest in a lineage, it *must* therefore confer a survival advantage. Evolutionary biologists use the latter kind of backward reasoning all the time. You can't say that the reason they survived is *because* they were the fittest, and also that the reason that they were the fittest was *because* they survived. It's circular.

So, if it's not the "fittest", what does evolution select for? A novel answer is given in Robert Wright's 2000 book, *Nonzero: The Logic of Human Destiny* [Wright 2000], echoed by many modern evolutionary theorists:

> *Evolution selects for positive-sum games.*

A *zero-sum* game is one where, whatever one player wins, another loses. A *positive-sum* game is one in which, one player's win can also result in a win for others. Wright uses the term *nonzero*. But here we'll use the term *positive-sum* since negative sums don't make much sense.

What positive-sum selection means is that if you compare two situations of interaction between organisms, one that is zero-sum, with another that is positive-sum, the positive-sum situation "wins". Here, winning means that the species involved will get an advantage in survival and reproduction, especially in the long run.

The situations which count as "survival of the fittest" are generally zero-sum situations. If there's a predator-prey relationship, then whenever the predator gets better at catching prey, the prey get worse at surviving. Similarly, if the prey get better at avoiding predators, the predators get worse at surviving. Zero sum.

Situations of symbiosis and other kinds of inter-organism cooperation are positive-sum situations. Many examples of symbiosis evolved from predator-prey relations, when the prey evolved an "if you can't beat 'em, join 'em" strategy. Wright also extends the same argument to evolution of the history of human societies via *memes*, ideas that evolve as they are transmitted socially.

Meme evolution follows a parallel path to gene evolution. A superficial reading of Darwin leads to the impression that he emphasizes the role of competition between organisms, especially through the process of natural selection. Those that are best equipped to outcompete their peers succeed in the competition for food, sexual partners and other resources. They are more apt to survive and reproduce, and therefore are the fittest.

Darwin was perhaps less appreciative of the complementary role played by cooperation between organisms. There are plenty of examples. Perhaps the greatest is multicellular organisms, including us. In the process of evolution, bacteria learned to form colonies to support each other's survival, and later, cooperate to form multicellular plants and animals. It is now thought that mitochondria, the energy generator for cells, evolved from independent bacteria invading cells, that joined forces with their hosts.

There are many examples of symbiosis between different species. The clownfish feeds on natural enemies of the sea anemone, which in turn provides waste products that feed the anemone. Startling new discoveries of the human microbiome show that we live in cooperation with huge numbers of microbes, especially in our intestines, that are essential to our ability to digest foods and for our immune system [Hattori 2009].

Cancer

Atkipis et al. [Atkipis 2015] provide fresh insight into the causes of cancer. It outlines five different mechanisms by which cells cooperate with each other, common to organisms throughout the evolutionary tree. It then hypothesizes that cancer is the result of a breakdown of one or more of those cooperative mechanisms.

Cancer occurs when cells reproduce uncontrollably. Normal cells have a mechanism of *programmed cell death* have been evolved to assure that cells die at the "right time" to promote survival of the larger organism of which they are a part. A mutation can cause the programmed cell death mechanism to malfunction. In the short term, the mutated cells proliferate wildly, so they're successful in creating a tumor, but in the long term they cause the death of the entire organism, including the cancer cells that started it.

The authors of this article don't mention the Prisoners' Dilemma, but let's recast it in those terms: Cells normally *cooperate* with one another to maintain a healthy organism (*Reward*), which paradoxically, includes the cooperation of dying at just the right time. A cell that has a genetic defect, instead, can *defect*. It can gain a short-term advantage in survival (*Temptation*) by not dying at the appropriate time, thereby creating a tumor. But in the long-term, it hastens the death of it and its cohort (*Punishment*). The cause of cancer is cells defecting against one another.

Finkel, Serrano and Blasko in *Nature* [Finkel 2007], hypothesized that bodily changes in animals due to aging are also a part of this cooperative mechanism. Since mutations accumulate as time goes on, as a cell gets older, the chances of it mutating, and turning cancerous, increase. So the cell slows down the reproduction rate to reduce the chances of cancer. Basically, the reason we get old is to decrease the chances that we might get cancer.

Group selection is a common occurrence in evolution. If a group of organisms are competing with each other, they're playing a zero-sum game. They'll waste resources competing and miss out on the opportunity cost of cooperation. They won't be as good at surviving and reproducing as groups of organisms that have evolved to cooperate with one another [Marean 2015]. Evolution uses the same algorithm for selecting amongst groups as it does for individuals.

Arguably, the reason humans have evolved big brains and language is to cooperate with each other and form societies. Not (or at least not only) to fend off sabre-tooth tigers or for an arms race for mates. And so cooperation has evolved in the human race.

So, it seems that evolution actually has two strategies: a competitive one, and a cooperative one. What influences whether nature takes a competitive or cooperative path?

Scarcity = Competition; Abundance = Cooperation

It's our contention that one of the major determinants of whether cooperation or competition dominates in evolution is the relative *scarcity vs. abundance* of resources. Scarcity encourages competition. Abundance encourages cooperation. Why?

Remember, the TRaPS Inequality is

Temptation > Reward > Punishment > Sucker's Loss

Scarcity, by definition, means there's not enough to go around. In evolutionary terms, that means, given the available resources, some organisms will live and others will die. In the worst case, then, getting the Sucker's Loss means: you die. Since some organisms will live, those most likely to live will be those who get the most, i.e. the Temptation. That's a powerful motivator to try to make sure you get the Temptation and not the Sucker's Loss. The suckers will get eliminated from the gene pool.

Now, depending upon exactly how bad the scarcity is, and what the needs of the organisms are, the line constituting the live-or-die boundary can occur at different levels of resources. Let's start by looking at the case where the scarcity is so bad that a single organism can't survive unless it gets the Temptation.

We'll represent that by showing Temptation, where the organism lives, with a smiley face. The rest of the outcomes, where the organism dies, have a skull-and-crossbones.

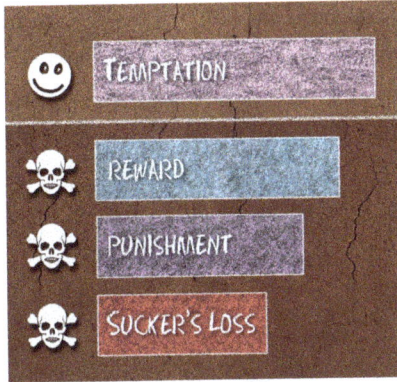

So the cooperate-or-defect choice the organism is faced with is:

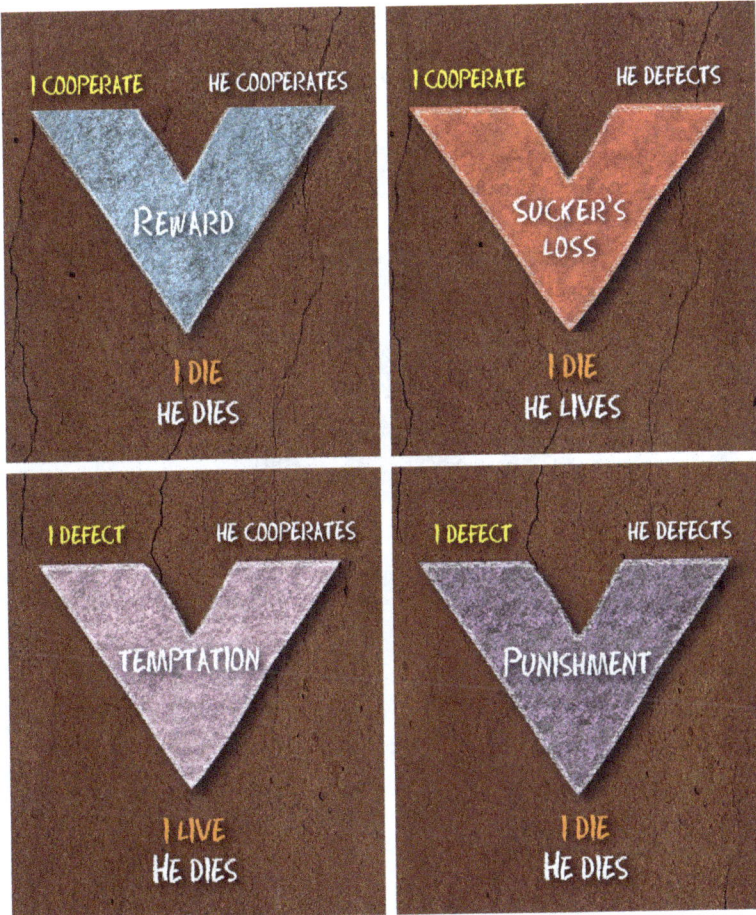

Then, the organism is forced to defect in the hope-against-hope that the other will play the Sucker and die as a result, thereby enabling the first one to survive. There's no point in trying to cooperate, because the best you could do would be the Reward. Not enough. Evolution will select for competition.

But from the group selection point of view, it's not great. Since we've posited that in this condition of scarcity, there aren't enough resources to ensure survival of all, and we've evolved the species for competition, an organism will likely defect. Then they get the Punishment, taking the chance that the entire species will go extinct. But in this case, the species has to take that risk. It's got no other choice. Them's the breaks when you've got scarcity.

Next is the case where both the Temptation and Reward levels permit survival, but the Punishment and Sucker's Loss don't.

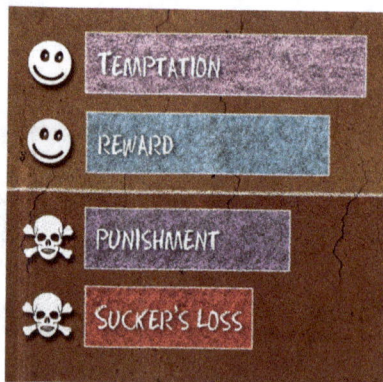

This scenario supports a mix of cooperation and competition. The choices are:

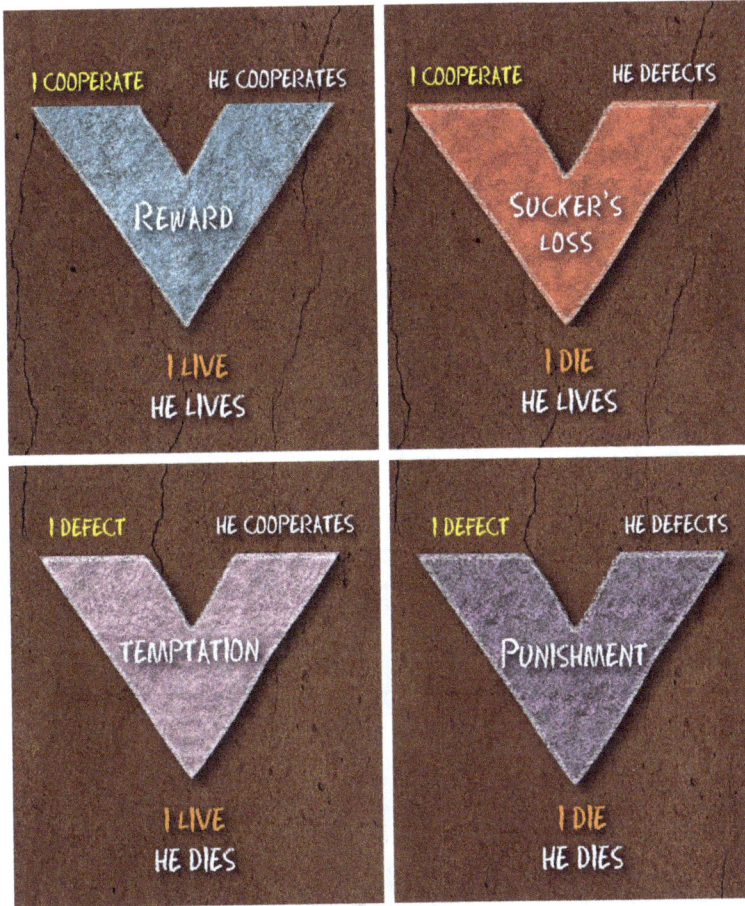

Mutual cooperation results in the Reward. There's no reason to take the further risk of defecting in order to try to get the Temptation rather than the Reward, since even the Reward is sufficient. But we still don't have quite enough for everybody, so somebody's got to play the Sucker. We can tolerate some mutual defection, their Punishment also joining the ranks of the losers. We still need at least some competition to determine who loses out, as long as that doesn't generate too many more losers than needed to account for the deficit implied by the scarcity. And certainly we don't need the level of competition that might risk extinction. That's much better from the group selection point of view.

Then, we've got the case where even the Punishment is enough resources to live on, and you'll only die if you get the Sucker's Loss.

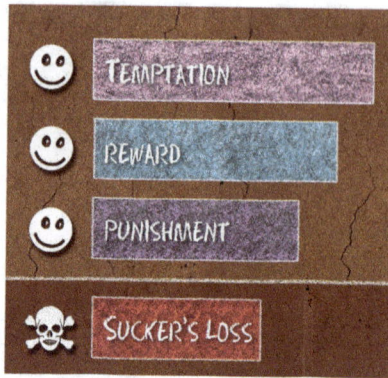

Again, no point in defecting, trying for the Temptation, if the Reward is enough. And we need fewer losers than before, therefore less competition, since the scarcity deficit is smaller. This shifts the tradeoff even further towards cooperation.

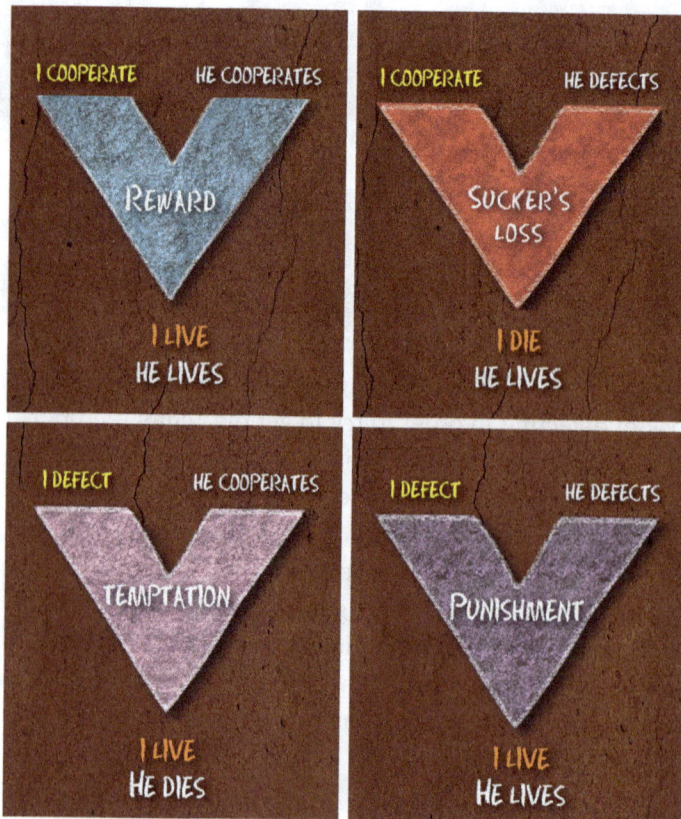

If all the levels are survivable, then we don't have scarcity at all, by definition. There's no natural selection pressure forcing competition.

Surplus resources can improve the organisms' lifestyle, invest in long-term security, and increase opportunities for reproduction. As we've seen in the classic Iterated Prisoner's Dilemma, cooperation is the best group strategy.

Sometimes, you hear arguments to the contrary. Some will argue that, instead, it's scarcity that forces cooperation. In tribal human societies, cooperation on a hunt is more important in times of scarcity than abundance. If they don't cooperate in the hunt, everyone will starve. There are slime molds that live as single-cell organisms in times of abundance, but form colonies in times of scarcity.

But these are not Prisoner's Dilemma situations. They don't satisfy the TRaPS Inequality, in particular the requirement that *Temptation > Reward*.

Imagine you're trying to decide whether to cooperate with a fellow tribe member on a hunt. If everything goes well—the hunt succeeds and you divide up the proceeds (*Reward*)—that's better than possible mutual starvation (*Punishment*). So we've got *Reward > Punishment*.

But there's a critical moment when you've just killed the animal together, you have to cooperate with your partner to divide up the meat. Your partner has to be trustworthy. Yes, you cooperated on the hunt. But what counts here is the entire process, which includes both the hunt and divvying up the spoils. If they're not trustworthy, they might think that the advantage of getting it all rather than just a share, is worth *defecting* (*Temptation > Reward*). In that case, you're out of luck. Remember, they've got a spear.

Scarcity makes this situation worse, as the difference between 100% of the meat and just 50% of the meat might make more of a difference in survival in times of scarcity, so it's more tempting to defect. They even might be more worried that the scarcity might cause *you* to defect, so they might be willing to take the chance that you'd both be disabled by the fight (*Punishment > Sucker's Loss*). Many's the caper movie where a thief and their partner pull off the heist, and one thief is gleefully counting the money when they turn around to face their partner's gun.

So, cooperating *with reliably trustworthy partners* still makes sense, perhaps even more so, in times of scarcity. But cooperating *with possibly untrustworthy partners* makes less sense the worse the scarcity is.

Mullinaithan and Shafir have written an excellent book on the psychology of *Scarcity* [Mullainathan 2013]. Scarcity has both benefits and disadvantages. The benefit is that it causes focus—it causes people to concentrate on what's most important, and ignore what doesn't require immediate attention. It encourages risk-taking that counters people's usual risk-averse bias.

The disadvantage is that focus causes people to be blindsided to peripheral threats and opportunities that might suddenly become important. For example, firefighters on the way to a fire are focused on getting to the fire as fast as possible, so they tend to forget to put on seat belts. The shocking statistic is that 25% of firefighter deaths are in traffic accidents on the way to the fire.

People who experience scarcity of money (poverty), time (chronically stressed people), or other essentials are less rational, in general make poorer decisions, and are less cooperative with each other. They defect more often in Prisoners' Dilemma experiments.

Over time, scarcity and abundance cause different kinds of feedback loops. Scarcity causes agents to waste resources on fighting with each other, which uses up resources that could have been used for reducing the scarcity in the first place. That's the "guns vs. butter" tradeoff. It's a negative feedback loop.

On the other hand, abundance permits agents to cooperate and the reward of cooperation leads to further riches. The surplus not necessary for immediate needs can be "invested"—to fund longer-term activities that might pay off sometime in the in the future. Or it can serve as "insurance" to protect against possible future scarcity. Both of these are likely to lead to more abundance in the future. This is a positive feedback loop.

We've already talked about how the Prisoner's Dilemma can be viewed as a short-term vs. long-term tradeoff. Especially in situations of scarcity, we can find ourselves on the knife-edge between cooperation and competition. The fact that each creates self-reinforcing feedback loops makes it all the more critical that we make the right choices.

Past scarcity and future abundance

So, if scarcity vs. abundance is a crucial factor, where is the human race? Historically, from the Stone Age to recent times, humanity has lived under conditions of scarcity. Even when hunting or early agriculture succeeded in providing abundance for a period of time, it usually was followed by a population explosion and/or wars that reduced conditions back to a basic level of scarcity. Malthus was pessimistic that humanity was doomed to remain in this condition. For most of humanity's history, no more than a small percentage of the population could be assured of having its basic life needs met in a reliable fashion.

Now, things are different. At present, we have a mixed situation that has pockets of first-world abundance and pockets of third-world poverty that are pretty much just as bad as the Stone Age. In general, as Pinker argues [Pinker 2011], things are getting better. We have relatively fewer wars and a fewer percentage of people are living in poverty than ever before.

This is largely due to technological innovation. Agricultural technology feeds people. Communication, computing, and collaboration technologies used in large organizations allow people to work together effectively to be productive and solve problems. Political techniques like modern representative democracy and markets (though we'll criticize them handily later) work better than earlier forms of anarchy, dictatorship, feudalism, and Communism. But this is all historically very recent.

Our societies have evolved with scarcity in mind. Our social and economic structures are designed for conditions of scarcity. As we have seen, that means that they have been set up assuming zero-sum competition. Because our educational systems teach that scarcity is inevitable, some people can't even imagine that it could be otherwise.

Our perception of reality is out of date. Much of humanity has changed from an environment of scarcity to one of abundance, but our self-knowledge hasn't changed. Education, like most other institutions is conservative. By updating our understanding of our situation, we can facilitate cooperation.

Because of the feedback loops, increasing cooperation will lead to abundance, making it easier to cooperate, and so on. It's our contention that now, the primary cause of war, poverty, and other societal ills is exactly the assumption that we'll always have war and poverty. Because of that, we get trapped in the negative feedback loop of defection.

If we can believe it's possible, or even likely, to have abundance, cooperation will both aid in achieving it, and also make it easier to cooperate. They key factor in bringing this about is educating people. That very change in mindset will itself launch the evolutionary process that will cause it to happen.

While there are still modern-day Malthusians, we believe that the march to abundance will not only continue, it will accelerate. We're computer scientists by trade, and we know how it will happen. The key enabling technology in software is Artificial Intelligence, and in hardware, personalized digital manufacturing. In subsequent chapters, we'll expand upon how these developments will result in the prospect of abundance.

For progress, we need our social institutions to transition from a scarcity mindset, encouraging competition, to an abundance mindset, encouraging cooperation.

Recap

Take a deep breath. We're just about done with the theory section of the book. Think about what you've learned.

You've got the basic conceptual tools of game theory for thinking about the tradeoff between cooperation and competition. You understand that evolution, especially under conditions of abundance, encourages cooperative, positive-sum situations.

We're now in a position to fully state the central argument of the book. Here it goes.

- Understanding the tradeoff between cooperation and competition is crucial to many important problems in economics, politics, social relations, and other areas. Neither cooperation nor competition is best under all circumstances.

- Many important situations can be described by the pattern of the Prisoner's Dilemma. If each agent considers just its own point of view in the short term, it will decide to defect (that is, to compete or choose not to cooperate). But from a more global, long-term view, the right decision is to cooperate, which leads to the best results for the group, and ultimately, for individuals.

- One important factor that controls which way the tradeoff goes, is the relative scarcity or abundance of resources. Scarcity encourages competition. Abundance encourages cooperation.

- Historically, with few exceptions, humanity has lived under scarcity.

- Due to information technology, especially AI and Makerism, we now have the possibility of replacing past scarcity with future abundance. Humanity is on the verge of having the capability to provide for everybody's physical needs. We will refer to this as post-scarcity.

- Unfortunately, as a result of historic scarcity, we have developed zero-sum social, economic, and political institutions. They are obsolete. We have traditionally prioritized competition over cooperation. That ends or we do.

- But improving technology alone won't get us to abundance automatically. If we keep our present zero-sum economic and political structures, even the "bigger pie" won't improve the lives of most of humanity. It won't be distributed evenly, as shown by the Ultimatum Game, and present inequality.

- We need to change our zero-sum, competitive mindset and institutions to positive-sum, cooperative attitudes and structures. This will both create abundance, and also make it easier to cooperate. We'll get on a positive feedback loop.

- Then we'll all live happily ever after.

Part 2
Does human nature allow us to get along?

Chapter 4
Is it even possible to get along?

Let's start by asking a fundamental question: How can you get people with two different opinions to agree? If I hold opinion X, and you hold opinion *Not X* (and in the case where there is no obvious way to "split the difference"), how is it at all even *possible* to come to an agreement?

The only *a priori* solutions might seem to be that either one or the other has to win out, based on who has the most power. Or that one of the participants has to change their mind, which is often difficult, because they may feel that betrays the values that caused them to take the position in the first place.

The problem is that many people think of a negotiation as being a competitive event, like a prize fight. They think the object is to have your own point of view prevail, by a show of superior strength over the other party. The goal is to land the devastating punch that the other party has no effective counter for.

In the competitive view, logical arguments are a weapon, like a clenched fist. Both parties are obliged to put up the front that the most logical argument will carry the day. But, in deciding what to say, they don't necessarily search for the most logical argument.

Instead, like lawyers do, they cherry-pick the best arguments to present their side, and the ones they feel the other side will have the most difficulty responding to. In the worst case, they use tricks that exploit people's cognitive biases and limited rationality, such as stacking the deck, or crass appeals to strong emotions.

If the participants are debating for the benefit of independent third parties who are trying to make up their minds, like politicians trying to influence voters, the third parties are just left with a game of Liar's Poker. They have little to trust or rely on to make up their minds.

Goal stacks

Don't stack the deck—merge your goal stacks. Imagine that your mindset in going into a discussion is not that it's "us vs them". Imagine that we don't even think that there are two sides. There's just *us*, and we're trying to collaborate to get what's best for all of us. We're all on the same side. (If you *have* to think of another side, because you think having another side will help motivate your team, let's imagine the "enemy" to be human suffering, poverty, unhappiness, etc. The disagreement itself is what we're all trying to fight.)

Differing opinions are typically not the whole story. Opinions that people have are not isolated from one another. There's a structure to all the opinions that a single person or group might have at any given moment.

It's determined by people's values and goals. People have reasons for believing in one thing or another. Some are more important than others. And they are typically arranged in a hierarchy of importance, in what AI people call a *goal stack*.

The goals on a stack are linked to each other by answers to the question "*Why?*". A small child might wear out their parents by asking, "Why are we in the car, Daddy?". Parent: "To go to the airport." "Why?" "To take a plane." "Why?" "To go on vacation." "Why?" "To have fun." Don't get mad at the kid. They're trying to learn your goal stack.

A common mistake in negotiation is to compare positions between the parties, notice where there are differences, and try to "hammer out" the differences one by one. As with any hammer, there's always the danger of hitting your thumb instead of the nail.

Instead, negotiations should begin by trying to understand what values and goals the parties have in common, and try to expand that to as wide a range as possible. Only then can you work on resolving differences. Common goals become criteria which both sides can use for evaluating candidate solutions.

The idea is to first work up the goal stack chain to the point where all parties share common goals, including the common goal of coming to terms with each other, and then work downwards to resolve differences (consistent with the shared goals) and secure mutual agreement.

Let's say we have a labor-management dispute where the workers want higher salaries and the management wants lower salaries. If that were all there were to it, the side with the most power would simply win. At best, they could split the difference.

But they both have a common interest in seeing a successful business with happy, healthy workers. Labor could make arguments that the business would lose the best workers to higher-paying companies. Or that the wages were difficult to live on, causing workers to take exhausting second jobs. Or that the company was doing well, and they deserve a fair share.

Management, for its side, could open the books to labor so that they could independently verify that if the wage increase were implemented, it might jeopardize the competitiveness of the business or cause layoffs. Or it could show a competitive analysis showing that wages were comparable to competing companies. Or either side could propose a profit-sharing plan that would automatically adjust to market conditions. Or propose arbitration or some other "fair" way to decide, since both sides share an interest in a fair decision process.

What if the goal stacks don't match?

The reason to be optimistic that agreement can almost always be achieved is that, since we're all human beings, the topmost levels of our goal stacks are quite similar.

Psychologist Abraham Maslow wrote about the *Hierarchy of Needs* [Maslow 1943] he felt were common to humanity. People first need their basic survival needs met: food, clothing, shelter. Then they need good social relationships, a feeling of security, and finally, what he called *self-actualization:* love, meaningful purpose to life and activity, and feeling of achievement.

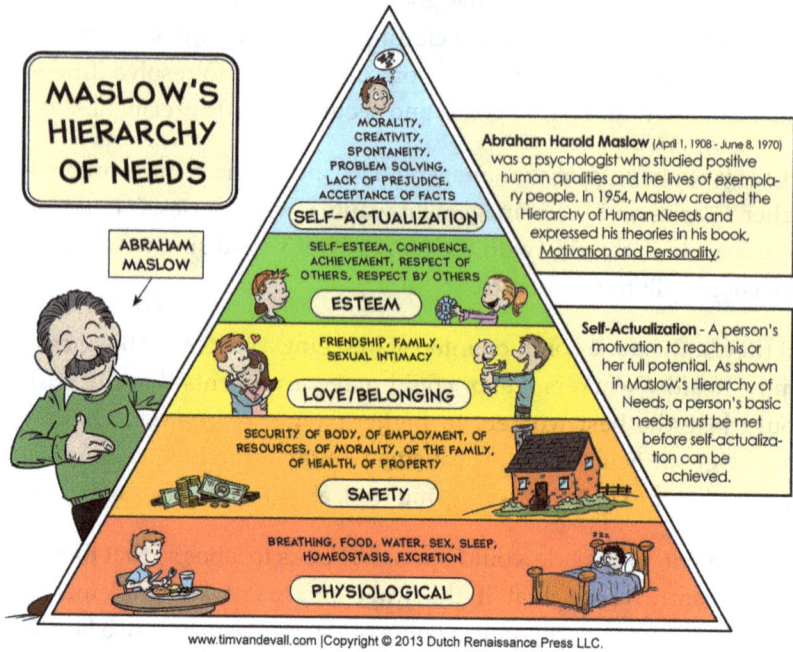

MASLOW'S HIERARCHY OF NEEDS

ABRAHAM MASLOW

MORALITY, CREATIVITY, SPONTANEITY, PROBLEM SOLVING, LACK OF PREJUDICE, ACCEPTANCE OF FACTS

SELF-ACTUALIZATION

SELF-ESTEEM, CONFIDENCE, ACHIEVEMENT, RESPECT OF OTHERS, RESPECT BY OTHERS

ESTEEM

FRIENDSHIP, FAMILY, SEXUAL INTIMACY

LOVE/BELONGING

SECURITY OF BODY, OF EMPLOYMENT, OF RESOURCES, OF MORALITY, OF THE FAMILY, OF HEALTH, OF PROPERTY

SAFETY

BREATHING, FOOD, WATER, SEX, SLEEP, HOMEOSTASIS, EXCRETION

PHYSIOLOGICAL

Abraham Harold Maslow (April 1, 1908 - June 8, 1970) was a psychologist who studied positive human qualities and the lives of exemplary people. In 1954, Maslow created the Hierarchy of Human Needs and expressed his theories in his book, Motivation and Personality.

Self-Actualization - A person's motivation to reach his or her full potential. As shown in Maslow's Hierarchy of Needs, a person's basic needs must be met before self-actualization can be achieved.

www.timvandevall.com |Copyright © 2013 Dutch Renaissance Press LLC.

That finding common values is almost always possible is practically guaranteed by the Maslow hierarchy. So it's a question of working our way up to the point where the parties share common values, then collaborating on how we can best promote the values we share in common, in a particular situation.

It's very important to understand the relationship between concrete positions on issues, and the values that underlie them. These can often be elicited by simply asking the other party, "Why do you hold that position?". Sometimes it may take several iterations of "Why?" before you get to what the other party really cares about. Rosenberg's *Nonviolent Communication* [Rosenberg 2003] shows the importance of eliciting and being sympathetic to the other party's needs, and presents concrete techniques for doing so. Eventually, you can build a conceptual model of the other side's goal stack and priorities.

Sometimes the positions can be quite different, but when you go up to the level of the values, it may turn out that both sides can agree on them. Sometimes the positions might seem unreasonable, but the values behind them are quite reasonable. Sometimes there might be another way of satisfying the values of both parties, that might involve concrete positions that will be more acceptable to the other side. Even in seemingly "intractable" situations, this can be the case.

It's always absurd to hear Arabs or Israelis insist that the conflict is so intractable that there's no possible solution. Of course there is. Most people in the world live in places where they do feel secure, despite neighboring peoples with differing religions or ethnic backgrounds. What could possibly be so special about these two groups of people that would preclude them from living in peace as others do?

There have been many other similar situations where longstanding conflicts have been resolved. Most dramatically, in Northern Ireland, where despite a centuries-old history of hostility, and terrorism in modern times, things now seem peaceful (August 2017). Similarly, in El Salvador, the Balkans, and other cases, peace was achieved. So why can't we solve this one?

If your goal stacks differ, trade

Some situations might be over-constrained, and the best you might be able to do in that situation might be a trade-off between mutually exclusive influences. But still, finding the best trade-off should be, a common value that both sides can work towards together. Here, the problem-solving mindset common in science and engineering, but sadly lacking in politics and business, is the best guarantee of success.

Scientists and engineers don't get upset at the prospect of a trade-off, they just set out to figure out what the best trade-off is, where possible, trying to quantify it. They figure out how to change the situation to make trade-offs less necessary or less painful in the future.

Uncovering differences in values and priorities can pave the way for creative solutions. If the parties value the same things differently, that can create the basis for a sensible trade. I give you the thing I care less about and you care more about, in exchange for getting from you the thing I care more about and you care less about.

People can change their mind on concrete positions when they understand that there might be different ways of satisfying the values that underlie those positions. They don't have to feel like they're surrendering or capitulating.

Back to our labor-management dispute, say the workers held the salary raise as a high goal. But they were also willing to make productivity improvements that didn't seem too onerous to them, but would make the company more money. Then, the employer might not mind giving the raise. However, the employer might not believe that higher salaries would raise productivity, so they make a deal that the amount of the raise would be tied to the amount of the improvement in productivity. Since the employees believe that their salary raises will increase productivity, they agree. Since the employer doesn't have to pay unless the improvements are realized, he agrees.

As in many of the situations we have been examining, this is another case where the principles of the Iterated Prisoners' Dilemma apply. A will to cooperate on the part of both players paves the way for a win-win outcome, where an attempt to compete, or to act on a fear of the other party, leads first to the danger, then the inevitability, of a poor outcome.

Once our goal stacks align, we can collaborate on solutions

Let's say a proposal is made (doesn't matter "which side" it comes from since we're all on the same side.) Now, we are going to discuss its merits and demerits. The very person who made this proposal should be able to say "This disadvantage of this proposal is it will decrease the health of population X". From the "competitive" perspective this person is inconsistent, because they are shooting down their *own* proposal. If they were a politician, they'd be accused of "flip-flopping".

But from the collaborative design perspective, they are helping everyone understand the implications of the proposal. Yes, they risk making it less likely that "their" proposal will be adopted wholesale.

But imagine that someone else says "Oh, now I get it and, yes, I see that that is a problem but we can mitigate it by doing Y". Now we have a *better* proposal than the initial one and the proposer should be delighted that the group has an even better solution.

The original proposer should feel great that they started off the conversation with a basically good idea, *and* they were big enough to articulate a misgiving which lead to an even *better* proposal. The "opponent" should also feel good about themselves. They understood the basic proposal, and a flaw, and then invented a fix for the flaw. This is *win-win*.

Let's imagine a variant of the above scenario. Instead of the original proposer coming up with the objection, someone else does. What happens? Well, often the original proponent will have a solution to this objection, they just failed to mention it in the original proposal. The critic did a great job. They *helped* the proposer better articulate the original vision.

Our third scenario is the same as #2 except that the original proposer, upon hearing the objection says "Whoops, I didn't think of that. I don't know how to fix it.". Had the critic not raised the objection, and the whole group adopted the original proposal, we'd have an easy consensus. But we'd only find out the flaw upon implementation.

Much better to find out earlier, before deployment. So the condition of having a plan with the known flaw is better than the original plan. Our critic has *added value*, and should be commended, even by the original proposer. Maybe a second objection will help us see a pattern to the objections, and assist in discovering a solution to both.

Reaching agreement by taming complexity

Complexity undermines our ability to reason clearly. The human mind has a limited capacity for complex concepts. This capacity is easily exceeded, especially when we're operating competitively, which tends to restrict thinking to "how to win". Thinking about collaboratively designing new alternatives helps expand the range of possible solutions. Furthermore, we can extend our cognitive capacity the same way we extend our physical limitations: with tools.

Computers are tools that fundamentally help us manage complex knowledge. Decision support software can help record and communicate the rationale for, and dependencies between, questions, positions, reasons, proposals, and their relationship with human values. It can help analyze and optimize trade-offs, and compute the combinatorics of possible deals that rely on value differences. We'll further explore this possibility in the chapter *Tools for Reasonocracy* [Ch. 27].

By cooperating with each other and with our machines, we can end the prize fights, and keep our eyes on the prize

Chapter 5
Interpersonal relations

When we decided to title this book, *Why Can't We All Just Get Along?*, we realized that one of the disadvantages of choosing that title would be that people might mistake it for a pop-psychology self-help book. It's not that book. Honestly, we didn't choose that name just to increase our paperback sales at airports.

How can we expect the US and Russia to get along, when we can't even be sure a bunch of college roommates won't strangle each other? Now, these situations aren't exactly the same, because the roommates have a personal relationship with each other, and governments don't. We're also not talking about the kind of individual, but impersonal, transaction a customer might have with a Starbucks barista. But any insights we can gain from improving personal relations might help in improving society.

Surely any attempt to get people to be more cooperative with one another has to boil down to individual relationships. No matter how much it may be in people's best interests to cooperate, if people don't like each other.... ain't gonna happen. No matter how much we present rational arguments for cooperation, if people don't feel good about cooperating.... ain't gonna happen. If we can't get along with each other, all the technological advances in the world won't result in a happy life. We can't ignore the emotional aspects.

We don't have a magic formula for getting uncooperative people to cooperate. Neither, at the present time, does psychology, psychiatry, sociology, or religion. Marriage counselors have a dismal success rate. Psychiatrists and psychologists don't seem to be able to effectively deal with a large percentage of mental health problems. The criminal justice and prison systems have a shameful recidivism rate. Religion seems as much the cause of conflict as the cure of conflict. Politics, for its part, does claim to have a technique for forcing cooperation—unfortunately, that technique is *fear*. So we need to search for a better one.

We still struggle with these issues in our personal lives. And we aren't so presumptuous as to suggest that information technology, where our expertise lies, is intrinsically any kind of solution to what are fundamentally people problems.

But some of the techniques of psychological and social sciences (and even religion, too) might help. There's good work in these areas, but like many other areas of research, even if the future is here, it isn't evenly distributed. Some genuine advances are quite recent. These fields are chronically cluttered with dirt, but that dirt hides diamonds. Because studded diamonds are easy to see, they're out front on the surface.

The basic problem is we don't know enough about how the human mind works. Especially its emotional and social aspects. We can't open the hood on the brain or download its software to see how it works. We can only surmise from very indirect experiments in biology or human behavior. These are pretty crude tools.

It's possible that sometime in the future, we'll know enough about human thinking, emotion and social relations that we will be able to "cure" hostile, violent, or sociopathic individuals with drugs, talk therapy or surgery. Such treatments will raise issues of autonomy. We still don't have a clear idea of what a normal or acceptable range of human behavior is, and debates rage.

We're still in the voodoo stage of understanding the human mind. Maybe someday, today's alchemy of understanding people will turn into chemistry. But for the moment, we're just trying to piece together what we can from psychology, cognitive science, neurobiology, and artificial intelligence.

Some people are just broken. Human thoughts and behavior are dependent to a large extent (we don't know how much) on the hardware—neuroanatomy and chemistry—of our brains. In severe psychiatric cases, antidepressants and anti-anxiety drugs have sometimes succeeded in ameliorating pathological behaviors where talk therapy and behavior modification techniques have failed. Maybe we'll discover that more problems have chemical causes, and develop drugs for them. If we discover neuro-anatomical causes of psychiatric disorders, perhaps some kind of high-tech "rewiring" brain surgery will fix them. Some disorders won't be curable at all, and coping strategies will seek to minimize harm to the patients themselves and others.

Artificial Intelligence can contribute to better understanding of the mind by making computational theories of human thought and emotion, and testing them in computational test beds that are independent of the biology of the human brain. Early experiments like Eliza and Parry [Weizenbaum 1966] explicitly tried to model social and emotional aspects of human interaction.

More recent books like Minksy's The Emotion Machine [Minsky 2006] postulate much more modern and nuanced theories of how the multiplicities of subcomponents of the human mind interact, giving rise to the vast array of human emotional behavior. It shows how emotion, far from being outside of, or opposed to, rational thought, is absolutely necessary to control and integrate our problem solving capabilities.

Picard's Affective Computing [Picard 2000], Mason's Emotion-Oriented Programming [Mason 2008], Cambria's emotional commonsense knowledge base [Cambria 2015], Liu's Emotion Buddy [Liu 2003], and other intriguing works aim to elucidate computational theories of affective aspects of human behavior. Advances in cognitive modeling of emotional states will inform educational and talk therapy models. We see some of this in movements like Cognitive Behavioral Therapy [Beck 1995], which try to identify negative beliefs that people with psychological problems become fixated on.

But aside from born sociopaths or others with serious medical conditions, society is full of instances of people failing to cooperate for emotional reasons, even when it would be in their best interests to do so and they are otherwise rational. Why do people act in uncooperative, aggressive, or overly competitive ways? Do people have an inborn need for aggression that no amount of intervention could overcome? We certainly hear people make such arguments when they argue for things like the inevitability of war.

In some circumstances, negative emotions really can be the right thing. Sometimes it's beneficial to be angry with truly harmful people in order to protect yourself (but probably in fewer situations than people think). Sometimes it's necessary to be assertive, to avoid group think. You can't be friends with everybody.

We argued that competition and scarcity have a mutually reinforcing relationship. Humans have uncooperative, competitive, aggressive behaviors, no doubt as a result of evolving at a time when scarcity was the norm. But in today's environment *we don't need the full amount of aggression our evolutionary heritage provides us* [Gibbons 2014]. Not only do we not need it, it hurts us.

One thing that the psychological literature seems clear on is that aggressive behavior is often motivated by insecurity and fear on the part of the perpetrator. One of our research projects was on the subject of combatting *cyberbullying*, online harassment of youth by their peers [Dinakar 2012]. In reading the literature on why children bully other children, blame was often placed on low self-esteem on the part of the bully. Fear of real or imagined threats, either physical or social, is often magnified far beyond anything actually justified by the situation.

If you dig into the history of bullies or criminals, you often find that they experienced some kind of trauma in their lives. Perhaps they had dysfunctional families, and in the worst cases, experienced violence or rape. People who experience violence get angry, and their anger often turns into violence against others, perpetuating a tragic negative feedback loop. Post-traumatic stress disorder (PTSD) amongst military veterans is an example of this. Psychologists have a maxim: *Hurt people hurt people.*

This violence-begets-violence dynamic is the same kind of feedback loop we see in the Prisoner's Dilemma. It is the *fear* of the other party defecting that leads one to choose defection. That, of course, is balanced against the *hope* that the other player will cooperate, which would lead you to choose cooperation. We're always on the knife edge between positive and negative feedback loops.

In Axelrod's initial iterated Prisoner's Dilemma simulations, Tit-for-Tat was one of the most successful strategies, and if players both cooperate, cooperation can continue. But two players playing Prisoner's Dilemma against each other with pure Tit-for-Tat strategies are unstable, because as soon as one player defects, the other will also defect, prompting another round of defection.

Techniques for getting people to be more cooperative encourage a positive outlook. Rosenberg [Rosenberg 2003] analyzes communication patterns between people, and shows that the implicit stances behind people's communication can either encourage or discourage cooperation. Becoming aware of these patterns and their consequences is a route to improving communication.

One of the books by the The Harvard Negotiation Project (source of the popular *Getting to Yes* [Fisher 2006] book series) explicitly deals with the emotional aspects of negotiation [Fisher 1991] and provides numerous practical techniques for managing the emotional aspects of discussions that may become contentious. Many other psychology, self-help, and conflict resolution books teach the same lessons.

Even techniques from Eastern traditions like meditation, mindfulness, and yoga, or meditative and prayer techniques from Western religions, can be helpful in combatting fear and insecurity, and lowering emotional tension in disagreements. They can get us to step back and contemplate our shared commonality as human beings and as being part of nature.

It's not necessary to believe in God or in the truth of a particular religious tradition to use meditative or mindfulness techniques. They can get us to concentrate on our good fortune in receiving the gifts of life and love, and teach us to empathize with our fellow human beings. In this context, disagreements, social status games, and past hurts come to seem small and insignificant. *We become more likely to bet on our hopes regarding other people than our fears.* Perhaps if we had meditation sessions before every business and political negotiation, the outcomes would be better.

Abundance to the rescue, sort of

Many interpersonal problems are more than money-deep. But we suspect that quite a number of personal conflicts can be solved by reducing scarcity.

One factor is density. The more people that live together, and the smaller space that they live in, the more likely it is that they have conflict.

Individuals need more space, and they need to live only a few at most per house. But that's often been unaffordable in the past, so interpersonal relations suffer.

Communes are few, we suspect because of too many conflicting personalities. A similar phenomena happens in companies of people banding together for making a business, though its less severe because the employees can escape at night. In either situation, though, when people must spend a lot of time with each other in close quarters due to economic necessity, they're wedged.

Abundance helps with happiness but only up to a certain amount. [Short 2014] suggests that the amount is highly dependent on cost of living in a town. The average household income in the USA for which "more money doesn't make you happier" is $75k/year.

Mental and physical health decrease with urban density according to [Recsei 2014]. We could not find studies supporting the hypothesis that happiness goes up with square feet per person so this hypothesis deserves study. Probably it too, like money, has a limit. We do observe that the 40-odd African countries are at the top of the "children per woman" list [CIA 2017] and at the bottom of the Happy Planet index [HappyPlanet 2017], though other reports do indicate that "big happy families" actually do exist.

If we solve scarcity, we will allow people to live with whom they want (or don't want), be that lots of people or few, resolving a lot of conflict. We need each other (at least for love, friendship, education, and entertainment but the "dosage" is crucial. Too little or too much, and it's bad. Abundance will allow us to adjust the dosage easily, thus benefiting interpersonal relations.

Abundance will improve interpersonal relationships in a lot of ways. Aside from sexual issues such as infidelity, one of the top causes of divorce is said to be money issues [IFDA 2013]. Many interpersonal problems are caused by competition for societal status, and when money becomes less of a determiner of status, conflict over these issues will be reduced. Those with mental health or emotional issues will be able to more easily get professional help if they can afford it.

Keeping up with the post-scarcity Joneses?

Some might say that competition for resources will simply be replaced by competition for social status. In that case, simply removing scarcity won't completely solve the problem, because people will invent new things to compete over.

It's another argument about what "human nature" really is. If you take a pessimistic view of human nature, you can believe that fighting and competition for social status are ingrained and unchangeable. It's easy to come up with things to fight over, even after people's basic needs are met. You can find things that are inherently rare (for example, singular art objects such as an original painting). If people want social status enough, they can arrange to fight over these objects. The most unnecessary is when the scarcity is artificially created around an imaginary construct, such as competition for honorary titles.

We agree that there are good evolutionary reasons why such competitive instincts developed. Mainly, they had to do with the conditions under which people evolved, primarily those of scarcity, with many zero-sum situations. It's worth asking the question, how much of these instincts are actually necessary and adaptive for the modern world? And even if they work well in the present, will the coming technological revolution that we're postulating change the game?

We've argued that abundance will "take the pressure off" scarcity-induced competition. But everybody knows examples of people who are well off, but manage to spend their time fighting over silly things. When that's the case, isn't it an example of something we should be seeking to change, rather than held up as a proof of the irredeemability of human nature?

Our competitive society selects for people with highly competitive personalities. It stands to reason that these people, even after they achieve material wealth, won't be able to turn off the personality traits that got them there in the first place. So they keep going. Many middle-class people believe that if they emulate the personality traits of the upper class, that will likewise be on the route to obtaining economic security. But that may not be a slam dunk. Steve Jobs may have been both a success and an asshole, but that doesn't mean that if you're an asshole you'll be a success.

Here's an analogy: let's think about the activity of *washing your hands*. Washing your hands frequently, is good for health. Most people ought to wash their hands more often than they actually do. But there's a limit. There are a small minority of people who feel compelled to wash their hands several times an hour, far more frequently than makes any kind of sense. We have a name for such people. They are victims of *obsessive-compulsive disorder* (OCD).

Money is a good thing. People need money to live. Most people could make good use of more money than they actually have. Again, though, there's a limit. People who have an insatiable desire to acquire more and more money, beyond any plausible personal need for it, as most CEO's have, should also be viewed as victims of obsessive-compulsive disorder. No wonder some rich people, if they let their guard down, confess that their material success sometimes leaves them feeling empty or unfulfilled.

So, no, we don't expect that, post-scarcity, competitive and hierarchical relationships will merely get re-instantiated over social status rather than material wealth. We expect social norms will shift away from rewarding and admiring hyper-aggressive people, towards favoring cooperation and personal expression. This of course won't be true in every single case or every single time, but there will be a historic shift in the balance. Keeping up with the Joneses is a fool's errand.

Conclusion

Despite all the forces that threaten to divide us—personal insecurity, fear of others, scarcity of resources for a decent life, competition for social status, past wrongs—we can learn to be resilient against problems that arise, and to adopt a positive attitude.

We don't yet know enough about the mind and about social relations to be able to reach every single person, cure every psychological ill, and deal positively with every social situation. Pinker [Pinker 2011] shows that, indeed, viewed at a large scale, the arc of history trends positive— we have fewer wars and violence and poverty than we have had at any time in the past, despite the unending flow of negative headlines from the news media (and recent terrorist incidents, which still don't negate the long term trend).

As psychology and other sciences improve, we'll understand more about specific techniques that will help people feel better about themselves and get along better with others. Though there are some known good techniques, and we expect more to be discovered, education is, as it is in most situations, crucial. We can't *all* just get along, until *each of us* learns how to get along.

Recap

- If we're going to have a new Age of Cooperation, we also have to think about individual person-to-person relationships, not just group relations in the economy and government.

- Psychology and related sciences still do not understand enough about the workings of the human mind and social relationships to be able to deal with all the reasons why people sometimes act aggressively or violently. Those emotions and behaviors are sometimes necessary, but we need to keep them in check.

- We fully expect that future advances in psychology, neuroscience, cognitive science and artificial intelligence will yield insights that will eventually enable a more scientific and effective approach to these problems.

- In the meantime, many techniques have been shown to be useful in setting contexts that put people in a more positive frame of mind, encourage cooperation, and discourage aggression. Meditation, mindfulness, and some aspects of religion can be helpful. Branches of psychology like Cognitive Behavioral Therapy, and Positive Psychology have shown promise.

- Future technology-based abundance will "take the pressure off" many people in dealing with problems caused by scarcity that exert psychological stress and cause fear.

- While aggressiveness may have been a legitimate response to conditions of our evolutionary past, it may now be maladaptive. We should be careful not to replace destructive competition for material resources with destructive competition for social status.

Chapter 6
Genderism

Many discussions about the issues of cooperation vs. competition eventually get around to issues of gender. (We prefer the term *genderism* to *sexism*. "Sex" refers to biology and an attention-getting activity; "gender" is a better term for the social construct, what we're talking about here. We may lapse occasionally into the conventional terminology).

Competition, it's said, is a "man's thing", and women are, by nature, more cooperative. So any debate about the relative merits of cooperation and competition is connected to the issue of male vs. female perspectives. This characterization is, in itself, controversial. Debates rage as to, first of all, whether it's really true that men are competitive and women cooperative. There are certainly examples of competitive women and cooperative men, or people taking different stances at different times. There's a range of attitudes in both genders, with considerable overlap.

Then, even if you agree with the idea of a gender difference, to what do we attribute that difference? Nature vs. nurture; genetics vs. environment? Is a cooperative stance selected for, due to women's role in childbearing, where cooperation between the mother and child, and between caregivers, confers a survival advantage? Or men's traditional role as a "provider", where historical scarcity has led to survival advantage for those who take a competitive attitude and are successful at it? Or simply the competition for desirable mates, which can explain these traits in animal evolution?

In the US, men tend to lean Republican (political Right) and women Democrat (political Left). We've heard the Republicans referred to as the "Daddy Party" and the Democrats the "Mommy Party". Men are more likely to advocate military measures like war (competitive) and women are more likely to advocate social safety nets (cooperative).

There's no denying, also, that historically, most societies have been male-dominated, and practiced sexism towards women. This is an explanation for why competitive attitudes and structures have dominated politics and economics throughout the centuries.

We've already seen an explanation for the persistence of sexism (as well as racism and other forms of discrimination) in terms of game theory, in the *Prisoner's Dilemma* chapter [Ch. 1]. Once you have a majority (or dominant) group, and a minority (oppressed) group, it can set up stable patterns.

Simulations [Hammond 2006] show that if the dominant group agrees to cooperate with other members of the dominant group and not with members of the oppressed group, the dominant group has an advantage that can form a relatively stable pattern. While perpetrators of sexism, racism, etc. try to pin their prejudice on some inherent characteristic like genetics, the simulations show that it is not necessary that the oppressed group have any kind of disadvantageous characteristic. All that it requires is that it be obvious to which a group a given individual belongs. However, the long-term cost of this is that the returns to both groups—including the dominant group—suffer as a result of the discrimination.

In the chapter *Intrinsic and Extrinsic Motivation* [Ch. 10], we discuss gender differences in motivation. The stereotype is that competitive activity is a motivating factor for men, while women may be more motivated by communicative and cooperative activity. Competition in economics, war, and sport is often cited as an incentive for motivating extraordinary effort. Teams in these areas are often cooperative amongst themselves, but count on drumming up competitive sprit against outsiders. That may not apply so much for the female perspective as for the male.

It's also obvious that gender roles in the modern world are evolving. More and more women are entering the workforce. More men are participating in childcare and domestic issues (though perhaps this trend isn't accelerating as fast as the former). Social strictures that tend to keep people in their respective gender roles are loosening. This is a good thing.

Could it be that increased participation of women in the economic and political spheres portends a shift towards cooperation in society? We certainly hope so.

But that shift won't be automatic. The few women who do make it to top positions in companies and government aren't always paragons of cooperation. Women like Margaret Thatcher and Indira Gandhi developed reputations as "iron ladies" and instigated wars. Of course, Thatcher and Gandhi reached their positions only by surviving a highly competitive, male-dominated process of selection as leaders. If we still have hierarchical, competitive, power-based structures, the gender of the top personnel may be only of limited impact. We diagnose the pathology of this leader-selection process in *No Leaders* [Ch. 22].

The inequality between men and women in intimate relationships is an age-old, persistent problem. Education in communication and cooperation skills will also help improve personal relations. Makerism will help women to become economically independent, permitting them not to have to enter into arranged marriages, prostitution, or more drastic sexual slavery. A Universal Basic Income is advocated by a number of economists. This would alleviate the inequity that work performed by women in housekeeping and child care, goes uncompensated in the current economy [Shulevitz 2016].

Experiments regarding collaboration in business and other problem-solving settings show that gender balance improves success [Hoogedorn 2013]. Other kinds of diversity, such as ethnic diversity, or balance between introverts and extroverts, also have similar effect. Note that in these cases, the diversity is amongst more-or-less equal members of a team, rather than the top leadership.

We need to take care when relating competition vs. cooperation to men vs. women. This would be wise on a scientific basis—we don't yet know enough scientifically about the causes and effects of gender differences to be able to say how much of what we observe in today's society is inherent.

While we're trying to discourage excessive competitiveness, that doesn't mean we should throw out the positive characteristics associated with masculinity, like courage, initiative, dedication and persistence. We maintain that cooperation and competition need to be in balance, but that the balance is presently skewed way too heavily in favor of competition. So, too, should masculinity and femininity, whatever they are, be in balance; the present balance is skewed way too far in favor of stereotypical masculinity, patriarchy, and sexism.

Masculinity and femininity are part of individuals' strongly-felt identity, and if they feel obligated to embody the stereotypical characteristics of their gender, that's limiting. Men sometimes feel they have to "act tough" at the risk of not being considered masculine enough; women sometimes feel unable to assert themselves for fear of transgressing their stereotypically feminine role.

Separating these issues would allow both men and women to be able to choose behaviors more appropriate to the situation than their stereotypes would dictate. Reducing genderism would go a long way to helping us all get along.

Chapter 7
Nationalism and racism are obsolete

It used to be that people didn't move around very much. If someone knew that you were born in a certain place, they could pretty accurately guess:

- Your skin color and ethnicity
- Your religion
- What language you spoke
- What food you ate
- What music you listened to
- Whom you married
- Lots of other cultural things

Since all of these things depended on geography, we could package up the idea of a homogeneous group in a well-defined geographic area, and call it a "nation". Governments were established independently in each region, because it was hard to cooperate with others if they didn't speak your language or understand your way of life.

But governments intent on increasing power over their citizens discovered that the divisions between people were useful for creating the *protection racket* known as "war". Wars are not clashes between cultures; they're an excuse for government leaders to extort resources from their citizens, and for the current leaders to fend off rivals to their power. (See the *Real Wars and Pretend Wars* section, in the *Jailbreaking* chapter [Ch. 1]. We examine the history that led to this state of affairs in the chapters *The beginning of history* [Ch. 19], *War* [Ch. 33], and *War crimes* [Ch. 32]).

Convincing the citizens that the outsiders are bad guys is an integral part of convincing the citizens that the fearless leaders are the good guys: "You'd better keep us in power if you know what's good for you". The first requirement to become a protection racket enforcer is to be able to easily tell the difference between outsiders and insiders. Ethnic, linguistic and cultural differences are especially useful for that. It's easy to generate fear of other cultures, taking advantage of people's natural inclination to fear the unknown.

It's a small step from there, to countries and borders. Nationalism has the effect of institutionalizing racism and cultural discrimination. Israelis and Palestinians, Sunnis and Shiites, Irish Catholics and Irish Protestants, are not "different peoples", as you sometimes hear. There is only one kind of person: *homo sapiens*.

Cultural differences do not imply "nationality"

This is not to deny that there are cultural, geographical and historical differences between groups of people. It's just that you shouldn't divide the world into "us" and "them" based on such aspects as language and customs.

Languages differ from one another, and it may be easier to say some particular thing in some particular language, but all languages get the job of communicating done.

Its tempting to think that people who don't speak the same language have a hard time communicating so they might have a hard time forming a coherent government. Note that India, a population of over 1 billion, has 10 languages that are spoken by over 25 million people each. We won't claim that India has a great government, but language does not seem an insurmountable obstacle. Now that AI is solving natural language translation, there's no reason each member of a congress couldn't have their own real-time language interpreter.

Geographic differences are often cited as another reason to divide people into nations. Desert vs. jungle vs. mountain climates might involve organizing agriculture or urban development differently, but that's hardly a sufficient reason for different governments. Different governments do allow some experimentation with alternatives for laws, but see the discussion in the *Some of Us* chapter [Ch. 8].

In today's world, cultural differences often do result in the inability of different groups to get along. But by simply dividing a country, we get international war instead of civil war, hardly an improvement.

We should also remember that a few hundred years ago, ancestors of our own society were quite guilty of many of the violent, oppressive, intolerant and dictatorial sins we now criticize in other cultures. We "grew out of them", as Martin Luther King's belief that "the arc of history bends towards justice" says. But others are behind the curve. Education is a better solver of such problems than fracturing a country.

United Nation

The original "Pledge Of Allegiance" for the United States, written in 1892, included: "one nation, indivisible, with liberty and justice for all". Its author had hoped it would be used by all nations. (http://www.ushistory. org/documents/pledge.htm) If all nations upheld the same ideals, then wouldn't it make sense to have just one united nation? We've outlined the advantages, but wouldn't there be disadvantages as well? The short answer is: not enough to outweigh the advantages.

Isn't the United Nations some kind of world government, thus proving that one world government can't work? Admittedly the UN is a kind of government ,and with over 7 decades of failure, it seems unlikely to ever succeed in fulfilling the most basic of its missions.

Its charter, signed in 1945, begins: "We, the peoples (sic) of the United Nations, determined (sic) to save succeeding generations from the scourge of war...". The UN sends ineffective "peace keepers" to war zones, but, more importantly, fails to stop wars in the first place.

Hint: it's not the "United" part, but the "Nations" part. Nations are fundamentally about power, and power loves war. The ambassadors to the UN are sent to New York by the most powerful leaders of each country, in an attempt to gain advantage over other nations.

The "emote and vote" process of the UN is the same power-based process as the governments of most of the democratic member nations. The more dictatorial, "plea and decree" governments, are even less likely to choose ambassadors that will cooperate.

Suppose, instead of modeling the UN after conventional countries, it was modeled after our proposal for government structure in the *Reasonocracy* chapter [Ch. 25]. Ambassadors would not be the emissaries of some power-based ruling elite, but rather, randomly chosen representatives of citizens. Ambassadors would not be skilled in the political art of deception to gain power, but in reason and cooperation, to achieve well being for all. The UN has failed, not because its charter has unrealistic goals, but because it doesn't have a process or participants that could possibly attain those goals.

American Unexceptionalism

John Perry Barlow expressed the obsolescence of nationality in his *Declaration of Independence of Cyberspace* [Barlow 1996]. By analogy with the US Declaration of Independence, Barlow identified the international community growing on the Net as a new kind of "nation". His declaration proposes sloughing off the limits of conventional, physical, national boundaries. It'll be a while before governments give up their prerogatives to divide people. Basically, though, he's right. Nations are a product of a power-based privileged aristocracy.

The USA is justifiably proud of its traditions of individual liberty, freedom of expression, commitment to justice, and economic innovation (though we express these imperfectly). But the USA as a nation doesn't have any patent on those ideas. The idea of "American Exceptionalism" is that those ideals somehow give us a right to a dominant position in the world as a nation, and the right to enforce our ideas of government upon others That's jingoistic claptrap. It makes others think we're arrogant and selfish.

Racism: A failure of science education

Racism is a failure of science education, but it's hardly ever talked about that way. The scientific question is:

> *What makes people the same, and what makes people different?*

The scientific answer is:

> *A hell of a lot makes them the same.*
> *Not a hell of a lot makes them different.*

What we call "race" in this society has no scientific basis whatsoever [Zimmer 2017].

The scientific concept isn't race, it's *genetics*. Take so-called "black people" and "white people". It's a well established scientific fact that humanity originated in Africa, so the original people were what we would call "black" today. "White" people are a mutation that occurred when humanity moved from Africa to Europe and Asia.

An explanation for the prevalence of the mutation is that it trades off resistance to malaria (enjoyed by black people, who needed it in Africa) for vitamin D production in the skin (needed by white people when they moved to Sweden, and had less sunshine). End of story. Says nothing about peoples' intelligence or personality characteristics.

Since skin color and ethnic features are inherited, you might think that black people are more closely genetically related to each other than they are to white people. Wrong. Genetic variation is a function of how old the lineage is. Africans have the most genetic variation because they have the longest lineage. For example, the tallest people in the world (Masai) and the shortest (Pygmy) are both Africans. White and Asian people have less genetic variation.

And on the face of it, the whole idea of "race" is just plain absurd. Why do we have "races" for people based on skin color, but not hair color or eye color? What's so important about skin, just that we have lots of it and it's easier to see? We don't have races for tall people and short people, even though that's another inherited characteristic that's easy to see. There are thousands of inherited phenotypes.

Dissociation of culture from geography

In the modern world, people are traveling, living in different places, marrying people of differing heritage. We won't get a homogenization of cultures, much as imperialist multinational corporations would like that to happen. What we're seeing is a *dissociation* of culture from geography and nationality.

We should make obsolete the motivation to immigrate based on fleeing war and economic hardship. Yes, solving scarcity will do that. Then, perhaps fewer people overall will emigrate, but those that do, will do so based on a desire to experience other places and cultures.

Cultures will hybridize and experience "hybrid vigor". Communities bonded by shared cultural interests can extend throughout the world, linked by the Net. Fans of rap music in Istanbul, Dakar, Marseilles or Jakarta create and enjoy an art form born in the Bronx and Compton.

From now on, almost every place will become multicultural. Perhaps we're biased, since we're from New York and LA, places which are already pretty much like that. With multiculturalism, we get a cornucopia of music, cuisine, and fashion. Most importantly, we get diversity of perspective and more knowledge resources to draw upon to solve problems. There's no point in fighting this trend. No point in trying to defend borders, build walls, stir up nationalistic sentiment.

We should enjoy the cultural diversity of humanity. There are enough examples of people getting along despite marked cultural differences, that we can at least be certain it's possible. Diversity shouldn't be an obstacle to getting along. Diversity should be something to celebrate.

Chapter 8
Can at least some of us get along?

In the modern world, organizations from families to governments are often dysfunctional. Is there any hope that, somehow, people could learn to cooperate in small groups? Is there, then, a possibility that we could scale that understanding up to larger groups?

There are plenty of examples of small groups that get contentious and disintegrate in a storm of infighting. Despite this, we observe that some small groups have a better track record of cooperation, without the large-scale, systematic exploitation found in industries and governments.

In small groups, people are more likely to know each other, and care for each other, as well as their community. It's easier to experiment in small groups, where changes can be made easily and rapidly without requiring large-scale political and economic movements. So we believe that small group cooperation is a fertile place to look for inspiration.

Here, too, technological change is facilitating cooperation in a way that hasn't happened before. The Internet is enabling small communities, perhaps distributed over a geographical distance, to connect synchronously and asynchronously. It enables experimentation with alternative governance structures. Computer mediated communication such as decision-support software enables people to solve problems more effectively than back-and-forth conversation or voting.

Consensus process

Numerous small groups eschew unstructured conversation, followed by voting, for a more productive process: *consensus*.

http://consensusdecisionmaking.org is a portal to the literature on the topic and also see [Butler 1987]. In ordinary terms, the word "consensus" is often used for a situation in which everybody agrees. Indeed, agreement, at least at some level, of all participants is the goal that consensus process strives for. Consensus process recognizes that it is not always possible to achieve unanimity on decisions. So it often strives to achieve a weaker form of agreement, one in which the vast majority do agree on an issue, but the dissenting minority agrees to "stand down".

That is, they register their "non-blocking" objection, but concede, for the sake of being able to arrive at a collective decision, not to stand in the way of the majority opinion. Conventional voting and majority rule remain as a last resort for truly intractable issues, but ideally, are rarely invoked.

Unlike a conventional democracy, the majority can't just celebrate their victory, then assume carte-blanche to do whatever they want. The majority has to hear out the concerns of the dissenters, and make a good-faith effort to address the concerns of the minority.

Blockers of a proposal will often block because they don't believe a proposal will, in practice, work out as well as its supporters think it will. In this situation, a potential blocker may set up a near term *milestone* to definitively test an aspect of the proposal. The other participants agree that if the milestone is not passed, the proposal will be withdrawn and/or another will be adopted. Because at each point in this milestone process we have agreement of all participants, it achieves consensus. This is consonant with the principles of scientific experimentation to test hypotheses.

In a voluntary organization, if the majority does not address concerns of the minority, the minority may walk. In US Democracy, where you can't walk away, minorities lose, period. If you have 49% of the population opposed, you upset nearly half the voters. That's pretty bad, especially considering we're all in the minority at least some of the time.

Work on consensus process revolves around designing procedures for people to use to discuss issues and come to collective decisions. One design criteria for these procedures is to make sure that they are inclusive, with all stakeholders having a voice in the issue, and everybody having a chance to contribute ideas. Care is taken to assure that no particular individuals will dominate. If everybody believes in advance that the procedures are fair, it increases the chances that the decisions that come out of that process will be accepted.

Many consensus processes begin with brainstorming sessions, that allow everybody to express their ideas for positive contributions to solve the problem, while reserving criticism and judgment for a later stage. Emphasis is placed on appreciating "win-win" proposals that benefit both sides of an issue, and "joint fact-finding", independent of people's positions [Susskind 1999].

This is important, because in most political discussions, people simply state their predefined positions, fail to listen adequately to other people's positions, and don't spend enough time searching for novel solutions that might be superior to their initial positions.

People do need some education and guidance as to how to interact with each other effectively in a consensus building process. The role of the *facilitator* in consensus meetings is crucial. The facilitator acts as a neutral third party to make sure everyone's voice is heard, to check that each side is understanding the others' positions and concerns, to cool hotheaded emotional displays, to encourage development of win-win proposals, and to keep everyone on track. Specific training in how to be a facilitator is essential.

Once a wide range of proposals have been put on the table, then discussion about choosing alternatives can begin. As in conventional political discourse, people can state the reasons for and against various proposals, and debate ensues. The focus of the discourse is supposed to be less on trying to convince opponents to accede to the original proposals, and more on inventing new alternatives and iteratively modifying proposals. This often takes the form of "friendly amendments" made to original proposals.

The goal is to take account of new opportunities and points of concern raised by dissenters, in order to garner wider support. That portion of the discussion can often get lengthy. But if everybody feels like they have a stake in a fundamentally fair process, they are more likely to accept the decision.

For consensus to be successful, the majority has to take the responsibility to meet the concerns of the minority. The minority has to take the responsibility of accepting the consequences of indecision or error if they decide to "block" the consensus decision.

Of course, it's difficult to make explicit rules that enforce the sense of responsibility and fairness that you need to make this process work. But at least consensus process has the aspiration of cooperation. This aspiration helps establish a social norm that leads to more productive interaction.

We believe that consensus process represents a fundamental advance over organizational systems such as Robert's Rules of Order, which enshrine a competitive process as much as the Marquis of Queensberry Rules (the rules for boxing) do [Susskind 2006]. Those rules may make boxing a "fair fight", but inevitably, somebody's still going to get hurt.

Consensus process differs markedly from the current state of US national politics (or for that matter, corporate politics). The basic stance is assumed to be competitive, both between candidates, and between political parties. Any behavior to the individual politician's advantage, such as presenting a one-sided view of an issue, or ad-hominem attacks on proponents of opposing viewpoints, is acceptable so long as it does not explicitly violate election laws and parliamentary rules. And even those laws do, in fact, get violated all the time.

The Harvard Project on Negotiation (PoN, www.pon.harvard.edu) and Consensus Building Institute (www.cbuilding.org) have a long history of work in translating the lessons of game theory for a general audience. They offer practical advice to business and political leaders aimed at encouraging win-win cooperation and defusing adversarial interactions.

PoN is best known for its series of popular books starting with *Getting to Yes* [Fisher, Ury, Patton 91]. These books provide sets of guidelines to be followed in meetings and other person-to-person communication. It tries to reframe the process of negotiation to a cooperative one where both sides jointly solve problems. It recommends trying to separate the problems from the people, and to always be mindful of improving the social relationships between the participants. It tries to move to mutual understanding of interests and values. It encourages coming up with creative solutions that make the pie bigger, rather than the zero sum competitive games where somebody has to lose.

When trade-offs do happen, it encourages trying to alleviate them, and establishing an objective basis for making them fairly. Susskind, McKarnen and Thomas-Lamar [Susskind 1999] have developed a comprehensive reference to practical consensus-oriented meeting techniques.

Other threads of work come to similar conclusions from alternative perspectives. The interest of the counterculture in forming intentional communities has led to the development of promising techniques [deTar 2013].

In 2011, the Occupy protests tried to construct a leaderless movement, with open meeting governance. Trying to do this without thoughtful methodology, and an untrained audience, in a high-pressure situation was so difficult that it didn't achieve much success. Part of the disaster of the Occupy "General Assembly" meetings happened because anybody wandering by could take up the group's air time, then leave. Speakers didn't even hear what others said earlier, and some merely wanted a platform to speak, without expending the time to listen to others. Working consensus groups should have a way external to the consensus process to restrict meeting participants. But Occupy did open some people's eyes to the possibility of nonhierarchical organizations and alternatives to voting.

CT Butler and Amy Rothstein's *On Conflict and Consensus* [Butler and Rothstein 87] is a handbook for conducting consensus meetings, emphasizing the formal process aspects, and used in a number of conflict resolution situations. Chapter 2 of [deTar 13] presents a good survey of counterculture consensus efforts, and some technical means to support them. Unfortunately, much of the counterculture work seems disconnected from relevant work in the business and government communities, like that of PoN.

Marshall Rosenberg's book, *Nonviolent Communication*, [Rosenberg 03] deals with the emotional aspects of communication in consensus and conflict situations. This is every bit as important as the procedural aspects covered by the books cited above. It deals with such issues as how to have empathy for other people, how to express both positive and negative feelings, how to express vulnerability, responsibility and guilt. It talks about how to ask for the things we need or want, how to apologize when we make mistakes. Most importantly, it instructs about how to listen effectively and non-judgmentally, and to appreciate the beauty in others. This is *not* the same as ordinary politeness, where the desire to make people feel better can actually get in the way of making good decisions.

Twelve-Step programs such as Alcoholics Anonymous and other support groups for dealing with substance abuse and other personal problems, have developed procedures for running their meetings and their organizations that also embody consensus principles. These groups are some of the most successful in helping people deal with the problems of their concern. They de-emphasize the role of "leaders" or experts, who are seen as performing a service role rather than a command role. Facilitators and participants take turns, assuring everyone is heard. Care is taken to welcome new members, who are assigned mentors from more experienced participants.

Lack of consensus on consensus

Consensus is not all roses. The good intentions of consensus are necessary, but not sufficient for optimal decision making. Although we're presenting our best understanding here, ironically, there's no complete consensus within the community, even on the definition of the word, "consensus" itself.

The current best practices for consensus limit the roles of voting and power politics, but they come with their own pitfalls. "Successful" consensus meetings will often end with everyone celebrating that they "reached consensus", regardless of whether the decision is good or not. This results in *least common denominator* solutions that receive the fewest objections. Politeness and over-concern with people's feelings may lead to avoiding difficult conversations.

Because the consensus process emphasizes agreement, it may put innovation at a disadvantage. Innovative proposals almost always have detractors simply because they are unexpected. Innovative proposals have to compete with the more generally accepted conventional solutions. A good facilitator can ensure that the consensus process doesn't become a rubber-stamp for the *groupthink* of conventional wisdom.

The reality of real-time

In both conventional decision-making, and in consensus processes, the primary venue for deliberation is the face-to-face group meeting. So they both share the disadvantages inherent in any real-time meeting.

Real time meetings tend to use one of two possible decision procedures. In one, everybody gets a chance to express their opinion (usually without much constraint on the expression, except for limited time). Then a vote is taken, and majority usually rules. We call that *emote and vote*.

In others, there's a designated authority who will make the decision, and the speakers aim is to exert influence on the boss rather than convince any dissenters. We call that *plea and decree*.

Neither provides the best opportunity for creatively solving problems and achieving consensus [Susskind and Cruikshank 06].

One move is to reduce reliance on real time, and move some processes online, where people can take their time to make considered responses, and see an overview of the issues rather than just react to the last thing said.

We also believe that technology will help facilitate the mechanics of procedures like consensus processes. [Lieberman and Fry 2013] describe Justify, an interactive decision-support tool that helps record rationale in an in-person meeting or online discussion. More about this in the *Tools for Reasonocracy* chapter [Ch. 27].

The real-time nature of meetings means that people often have difficulty remembering who said what, what was addressed, and why decisions were made. In short, meetings are too complex for an individual to understand in real time.

Justify can help with recording rationale in a structured fashion, helping people who join mid-discussion to catch up. It can decouple decisions from the pressure of real-time response, personalities and emotions.

Decision support tools can help with exploring the consequences of what-if scenarios and cooperation-competition tradeoffs crucial to making good decisions about complex issues. And if it isn't a complex issue, why involve a group in the first place?

Consensus in intentional communities

Both of the authors are children of the 1960s. Back then, frustration with mainstream society launched many experimental small group collaborations, such as hippie communes, food co-ops, and worker collectives. These emphasized democratic, egalitarian decision-making. Most of them failed, often due to personal infighting between the participants.

Groups did not have explicit procedures for dealing with disagreement that were resilient against personality conflicts. Aggressive individuals, acting as classic defectors, could easily disrupt the harmony of the entire group. So many of them failed, that the political Right uses that outcome as proof that any attempt at egalitarian governance is doomed to fail. They argue that hierarchical, command-and-control structures are "necessary to get things done".

But egalitarian communities didn't all die. Even today, some of the rural, "back to the land" communes are still in existence, after decades, such as Twin Oaks in Virginia (http://twinoaks.org) and Sandhill Farm (http://sandhillfarm.org) in Missouri. Two organizations, the Fellowship for Intentional Community (http://ic.org) and the Federation of Egalitarian Communities (http://www.thefec.org), act as a portal to such communities. Some of them have developed quite sophisticated governance and conflict resolution strategies that are responsible for their longevity.

And in urban areas, there are countless cooperative houses and apartments, sometimes called *cohousing* (http://www.cohousing.org) where residents share expenses, cook together, and cooperatively organize home maintenance. (I (Lieberman) lived in one for fifteen years.)

Many affinity groups even at a large-scale, operate with relatively nonhierarchical structures. Burning Man (since 1986), and the Rainbow Gathering (since 1972), are organized for temporary events that build up and operate the equivalent of a city for a few days at a time. Dance New England (http://dne.org) (since the 1970s) operates both ongoing participatory dances and performances, and an annual summer camp run completely by volunteers, consensus meetings, and representative committees.

Many academic and nonprofit organizations run similarly. Examples abound of small-scale volunteer groups that run mainly via consensus amongst equal participants, with representatives only necessary to take advantage of expertise, or to deal with matters that don't deserve attention of the wider group. In these groups, you don't see much of the circus of politicking, political parties, factions, power-hungry ambition, lobbying, etc. that appear in governments and corporations.

What a lot of these organizations have in common is some kind of statement of principles, mission statement, or governance document that emphasizes the cooperative nature of participation. They establish, by example, the social norm that people are assumed to be acting with the common interest at heart, even when they express strong opinions that might incite controversy. These social norms provide a basis for calling out occasional behavior that might be self-serving, or disrespectful to other participants, before it threatens to disrupt the harmony of the group.

The best example of an intentional community that operates largely on consensus is the *scientific community*, of which your authors are a part. In *The process of Science* [Ch. 24], we'll explore in depth how consensus operates in the scientific community, and how the social processes of science might point the way towards developing better cooperation models for government and industry.

Different strokes for different folks

Above, we've described intentional communities based on values like egalitarianism. But what if some group of people have different ideas about what their shared values should be? You could also consider groups like the Puritans as an intentional community. They emigrated from England to America because their values were different than the surrounding culture. They certainly weren't egalitarian.

One claimed benefit of "states rights" in the US Constitution is that different states can try out different rules, and we can see how they work out. Other states can adopt just those rules that are, overall, beneficial. In the world we have examples of Communism, Democracy, Oligarchy, etc. However, it seems like the world hasn't been so great, so far, at learning which systems work better than which others.

Geographic jurisdictions are a rather crude way to conduct experiments. Modern science does much better with controlled studies on select populations. It can choose much smaller and more targeted groups for conducting such experiments, making conclusions less ambiguous and more likely to lead to improvements of the experimental rules.

Let's take the highly contentious issue of gun control. Alaska might argue that guns protect citizens from grizzly bears. LA might argue that, guns in cars turn road rage into death. Some in LA might claim they need protection, not from bears, but from other people, but non-lethal weapons might fill that bill (See the *Guns* chapter [Ch. 31]). So, let's conditionalize the rule, not on states, but on the criterion that makes sense: No guns in areas with more than, say, 200 residents per square mile. Both urban and rural residents might not care if the rules are different in some other place.

There may, though, be a fundamental limit to how tolerant we can be of diversity. If a society or religion advocates violence, or oppressing a subgroup, there might come a point where the rights of the victims take precedence over cultural diversity. Americans are appalled by what they consider the oppression of women in some Muslim cultures. But some Muslim women in these cultures have written, "We don't consider ourselves oppressed, and who the hell are you to decide?". These are disputed cases, and difficult ones.

We do like the idea of having a diversity of intentional communities based on alternative value systems and mechanisms for organizing society. To prevent them from becoming oppressive or cultish, though, we recommend that one universal rule should be that anyone can leave at any time.

Cooperative enterprise

Once we understand that there is substantial benefit in exploiting untapped opportunities for cooperation, how do we put it into practice in the economy, especially when the dominant economy assumes a competitive stance? The danger is that small numbers of cooperators can be defeated by defection from a larger group of competitors, a danger that is often borne out by observation of Iterated Prisoner's Dilemma simulations.

One long-standing answer is the formation of economic cooperatives. In the long term, we believe that the new economy of *Makerism* will leapfrog the need for most large-scale economic organizations, including cooperatives. In the meantime, though, it's a great way of experimenting with more cooperative economic structures in today's world.

I (Lieberman) have my bank account and my mortgage in a cooperative bank (credit union), buy my food from a food co-op, shop at a cooperative university bookstore, have lived in housing cooperatives for decades, have bought car insurance from a cooperative, and have my bike fixed at a cooperative repair shop (which also offers to teach me how to fix it myself).

Fortunately, our society permits nonprofit cooperatives, though they still face discrimination from conventional economic institutions. Cooperatives provide viable alternatives to competitive economic institutions. They can be started on a small scale and grown incrementally. They don't require violent revolutions or mass protest movements, and can co-exist with capitalist institutions. The Internet itself is perhaps the best and most impactful example of a cooperative. It took over from for-profit, competing information services (anyone remember Compuserve and The Source?).

David Ellerman, a former economist for the World Bank, has put together a comprehensive book on the principles and history of worker cooperatives [Ellerman 1997]. His best example is the Mondragon cooperative in northern Spain (http://www.mondragon-corporation. com), which involves over 70,000 people in a variety of businesses, and has been operating since the 1950s.

Ellerman debunks the myth that large-scale enterprise requires absentee investment and hierarchical management. He shows how to structure the organization to fairly share profits amongst workers, with *internal capital accounts*. These accounts represent each worker's share of the ownership and profit rights of the firm, independent of the work-for-hire nature of the salary they receive.

The so-called "New Economy" enabled by the Internet refers to the fact that coordination can increasingly be provided by distributed computing. We can cut out the very expensive middlemen, called *disintermediation*. In the Ultimatum chapter [Ch. 2], we've shown how adding middlemen can drastically decrease the share of income received by producers later in the chain. In some cases, we don't completely eliminate all middlemen, but new Internet-enabled intermediaries can be more efficient and less costly than traditional business intermediaries.

Hidalgo [Hidalgo 2015] develops a more general theory called *Topocracy* that shows to what extent a producer is rewarded for the production itself, versus simply by virtue of their location in a network. All this points to disintermediation as a powerful agent of change.

Examples of disintermediation abound. Travel agents were disintermediated by airline reservation sites. Amazon disintermediated bookstores. YouTube disintermediated television stations. AirBnB disintermediated hotels. Zipcar, Uber, and bike share disintermediated transportation. The evolution of 3D printers may disintermediate most manufacturing companies.

Disintermediation represents an enormous opportunity to improve the overall efficiency of the economy. By that, we mean that middlemen don't take a cut, or their cut is reduced. The producer gets a higher percentage of the final sale price and the consumer pays less.

But there are pitfalls. The new for-profit intermediaries will have to resist the temptation (predicted by game theory) to become monopolists themselves (are you listening, Amazon?). And disintermediation, like any form of automation in a capitalist society, may reduce the number of "jobs". We need new mechanisms to make sure that innocent people do not suffer as a result, because under Capitalism, they surely are suffering.

In general, society would do well to encourage the formation of cooperatives and low-overhead private intermediaries. Traditional capitalist companies will undoubtedly take unfair defensive actions to put obstacles in the path of new institutions (via regulation, bribery, cartels, etc.). And we should put a priority on the development of new technologies that will facilitate disintermediation, such as automatic matching of supply and demand. Do-it-yourself products and services eliminate advertising, distribution costs and other unproductive overheads.

Small-scale cooperative decision-making is already here—in intentional communities, in affinity organizations, in the scientific community, in economic cooperatives. It's worth asking the question, what are these groups doing right?

Chapter 9
Sometimes, irrationality doesn't make sense

Some might find our effort to get people to act more rationally in discussion, argumentation, and negotiation a bit too idealistic.

People don't act rationally, they say. They choose their positions based on "gut feeling", emotion, social relations and power relations; not logical calculation and reasoned consideration. Logic, reasoning and argumentation are just used as tools or tricks to convince (or fool) others, in service of emotional or selfish goals.

It's remarkable how different professions seem to have radically different views of human nature. Classical mathematical game theory, argumentation theory, and mathematical logic view people as always acting to maximize self-interest. They posit a "utility function" that puts a specific number on every situation that says whether a specific individual will find that situation good or bad. This is a completely "rational" view.

Economists, also, view people as an idealized "homo economicus"— always acting rationally to maximize economic gain. But they don't have any idea of what actually provides value for people, except to quantify it in terms of money. Economists go so far as to work backwards, *assuming* that value is anything that people are willing to pay money for. If people are irrational and pay money for worthless products such as "pet rocks", economists assume that it is because they *must* provide some sort of value. That kind of reasoning is circular.

Mathematical theory completely ignores social, emotional, and aesthetic motivations ([Goleman 1995], [Mason 2008]) and others show that we have considerable "emotional intelligence" that is essential to human problem solving. Modern social network theories which analyze the graphs of social connections and flow of information between them show the importance of social network connections and patterns of collaboration in problem solving ([Hidalgo 2015], [Watts 2004]). Nobody yet has any credible theory of the role of aesthetics in problem solving.

Advertisers, marketers, media executives, and cynical politicians and business people treat people as being almost completely irrational (we'll refer to these people as *irrationalists*). They acknowledge people as being motivated only by a few primal animal impulses: need for food, shelter, clothing, sex, and power. They dismiss with disdain any suggestion that people form beliefs by conscious thought, or are capable of rational reasoning or morality. They seek to manipulate the public with emotional and subconscious appeals, and project a false sense of urgency in the hope that they can sell them a product before any rational thought has a chance to kick in.

Oddly enough, though, they view *themselves* as being rational in the economic sense described in the previous paragraph, using their pseudoscientific psychology to maximize their own economic gain. There's a galling sense of superiority amongst these people: they believe they are the shepherds, and we, the public, are the sheep.

The irrationalists will tell you that the proof of their claim is that it works. Advertising, bad as it is, sells products, they say. P.T. Barnum, perhaps the best exponent of this theory, said "Nobody ever lost money by underestimating the intelligence of the American public". He said some other things, too.

But be suspicious of any claims by the irrationalists. Remember who it's coming from and what they actually think of you. "Marketing science" will tell you that, in a lab experiment, given a choice between a row of no-name detergents and the heavily advertised Tide^tm detergent, people will pay more for Tide and choose it more often, without any rational evidence that it's any better. Maybe.

But that doesn't mean that one more TV ad for Coke^tm or Pepsi^tm in the already ad-saturated real world will have any real effect on its market share. Remember, the job of the advertiser is not to sell the product to the consumer, it is to sell advertising to the vendor. *The advertiser attempts to hoodwink the vendor, the same way they attempt to hoodwink you, the customer.*

Marketing studies that claim the efficacy of subliminal and emotional advertising are like the old toothpaste ads that had a white-coated "scientist" shill tell you their study proved that the toothpaste led to 44% fewer cavities. We predict that the fraudulent nature of most advertising will eventually catch up with it.

We predict that the advertising industry will experience a catastrophic meltdown very similar to what is happening with the so-called "music industry". The music industry's job was supposed to be to both increase listeners' access to music they love, and to provide a living for musicians. But decades of the music industry exploiting their "leverage", with low royalty rates for musicians and contractual restrictions, and exploiting the listeners with artificially high prices and limited availability, meant that the music industry was mainly working for itself. Once distribution alternatives were available, nobody had much sympathy for the music industry. It wasn't adding value to the product. Thus it collapsed. Similarly, advertisers exploit consumers and vendors alike. It doesn't add value to the product. Their turn is next.

Advertising is an example of an *arms race*. In the *Prisoner's Dilemma chapter* [Ch. 1] we discuss advertising as an example, and show how an arms race requires cooperation from people other than those who perpetrate it [Thompson 2014]. The continuance of war depends on people electing warmongering politicians, people paying war taxes, people joining the army, etc. If they stop cooperating with the arms race, it collapses.

To the extent that people are rational, our rationality certainly has natural limits. We only have so much time and energy. Several authors ([Ariely 2008], [Kahneman 2011], [Gigerenzer 1999], [Mullainathan 2013]) have detailed the extent to which our rationality can fail, usually due to one or more of the following causes: scarcity of time, effort, and resources; hidden assumptions; heuristics that are usually reliable but fail in particular circumstances; deliberation/action tradeoffs, etc. This is studied in AI as *cognitive biases* and *bounded rationality* ([Simon 1957], [Minsky 1988], [Manoogian 2016]).

There's been a recent movement to turn the deceptive tricks of advertisers and marketers around for good instead of evil [Thaler 2009]. This movement recognizes that people will inevitably have cognitive biases and limited rationality, but seeks to harness these techniques in service of positive instead of negative principles: tricking people into saving rather than spending; nudging people into exercising instead of watching TV; promoting choosing healthy foods instead of junk foods. They use gamification, incentive design, nudging and other techniques. If it works, and people don't mind or even like it, hard to argue with that.

Trouble is, even though these may have some effect, it's usually limited to extrinsic motivation in the short term (which is suitably convenient to be verified by short term experiments). We'll discuss more about intrinsic vs. extrinsic motivation, and short-term vs. long-term, later. This movement shares with the advertisers the basic disrespectful view of the public as sheep to be manipulated by the elite gamifiers and incentive designers.

Despite what the marketers will tell you, irrationality has its limits too. This was best expressed by Abraham Lincoln: "You can fool some of the people all of the time, and all of the people some of the time, but you can't fool all of the people all of the time". After a while, people's rationality will indeed kick in. People will begin to think for themselves rather than blindly accept what they've been told. Word will get around. The short term will turn into the long term. Push will come to shove. Something will hit the fan.

But this isn't assured. A problem with the Lincoln quote is, as advertisers have discovered, that it's possible to fool enough of the people enough of the time to make a profit. But if people are better educated, through media literacy, to think more critically about what advertisers are saying and why they're saying it, they may become less gullible. Once, ads were the only way you had of discovering the availability of products. Now, review web sites, recommender systems, online forums and other tools give consumers alternative ways of evaluating products. If people come to rely more on these, advertising will lose its bite.

Actually, we're all a mix of rational reasoning and irrational impulses. But the mix is complicated, it's different for different people, and it shifts dynamically and according to context. Neither psychology, nor any other science has yet figured out exactly what that balance is, nor how it is affected by external conditions. Given that neither the total rationalists nor the total irrationalists are correct, where are we? Where should we be?

We argue that the balance in contemporary society is far too much in the direction of the irrational. While acknowledging that people are not always rational, public discourse should probably put greater emphasis on appeals to people's rationality, than try to optimize utilization of people's irrational impulses. Though being rational is more work, and people are sometimes lazy, we ought to encourage people to do that work when appropriate, rather than appeal to their laziness.

A second difficulty is that being rational, or understanding a rational argument, requires basic principles of math and science. Here we run up against the poor state of math and science education in this country. Appeals to simple emotional drives don't require any of this.

Efforts to improve math and science education will reap enormous long-term benefits in improving peoples' ability to make better decisions across a wide variety of situations in their lives. They will be less susceptible to exploitation and trickery, to short-term impulsive decisions that will have deleterious long-term consequences. It's always better to give people the respect they deserve, and hope they'll live up to it, than disrespect them at the outset on the grounds they can't possibly merit respectful treatment.

Education now spends much time on rote memorization of facts. Instead, these topics need to be explicitly taught in schools:

- *Making a scientific case.* What counts as evidence. How to infer from data and specific examples to general principles, and vice versa. Scientific arguments are usually presented in science classes, but the *process* of scientific investigation itself receives little attention. Learning that is more important than any specific scientific topic.

- *Argumention and debate.* How to understand other peoples' points of view. How to convince somebody of something. How to recognize and deflect *ad hominem* attacks and other "debate tricks". Some of this is taught to debating teams, but it needs to be divorced from the counterproductive competitive stance of such teams. We also think that the technology of decision-support systems should be taught routinely, just as it's now unthinkable to train an accountant without using spreadsheets.

- *Conflict resolution techniques.* Finding common values and win-win solutions. How to handle anger and the emotional aspects of conflicts. How to disagree respectfully. Negotiation techniques. How to de-escalate conflicts.

- *Cognitive biases.* Advertisers, marketers, and politicians have learned to selfishly profit from the bounded rationality and unavoidable limits on rationality that Ariely, Kahnemann [Ariely 2008] and others point out. Students need to recognize and learn such tricks and understand how to think in spite of them. When you see the $3.99 product, immediately round up to $4 (maybe even $4.50 with tax or other add-ons you're supposed to forget about!).

Trying to work on improving rationality at least holds the possibility that we could work together to come to the best solutions possible in a given circumstance. All the irrationalists can promise is an endless war between competing impulses, competing nudges, competing incentives, and subliminal suggestions. Dan Ariely has argued that people are "predictably irrational". We say, they're both unpredictably irrational and unpredictably rational. Given that unpredictability, let's give them the benefit of the doubt.

Chapter 10
Intrinsic and extrinsic motivation

Take two groups of kids. Put each group in a room for an hour with paper and crayons for drawing, and tell the kids that they can spend time drawing if they want to, but they don't have to. What will happen? Since most kids like to draw, most will do at least a few drawings; the more artistic ones may draw a lot, some might not draw at all.

Now, say to the second group that you'll give them a dollar for each drawing they produce. What will happen then? Suddenly, you'll have drawings coming out of your ears—way more than what you got from the first group. What a dramatic demonstration of the power of incentive!

But come back a few weeks later, and something curious happens. Put them in the rooms again with paper and crayons, but this time, don't offer a reward to either group. What happens now?

From the initially unrewarded group, there won't be much change from what you got the first time. But the group that got rewarded on the first round, surprisingly, will likely produce many fewer drawings. Not just fewer than they produced the first time, but even fewer than the group that was not rewarded at all!

This is surprising, because you'd expect both groups to have a similar inherent level of interest in art. The disappearance of the reward might remove the additional incentive to produce, but it's worse than that. Giving people an incentive and then removing it actually destroys their motivation and appreciation of the activity itself. The experiments are reported in [Lepper 1973].

We'll bet, also, that the quality of the drawings in the rewarded group suffered, as well. Once the kids understood the game of reward–for–

drawing, they probably tried to optimize their return by making cruder and cruder drawings faster and faster.

This experiment shows the psychological difference between *intrinsic motivation* and *extrinsic motivation*. Intrinsic motivation for an activity means that you want to do the activity for its own sake. The activity is its own reward. You listen to music because you enjoy hearing it. You draw because you like drawing as a means of self-expression, even if nobody's paying you. Extrinsic motivation is provided by incentives that are external to the activity itself: rewards, prizes, grades, and rankings.

Intrinsic and extrinsic motivation aren't completely separate. You can have both some intrinsic and some extrinsic reasons for doing something. Somebody can try to make you do it, even if you want to do it anyway.

It's often overestimated how much external, extrinsic rewards motivate people to perform an activity. What rewards do is motivate people to obtain the reward. It's only because the reward-issuing authority links the activity to the reward that it encourages people to perform the activity. And that linkage is fragile—if anything disrupts the connection between activity and reward, the reward's power to motivate the activity disappears.

It's an admission by the authority that the activity isn't worth doing for its own sake, that the reward is necessary, otherwise people won't do it. If people come to believe this, then taking away the reward will leave people feeling like the activity isn't worth doing. Extrinsic incentives can actually discourage intrinsic motivation.

And, as the drawing-quality issue shows, the reward-seekers actually have an incentive to do the least possible to obtain the reward, launching a never-ending arms race trying to "game the rules of the game". We've already talked about the folly of trying to win an arms race, in the *Prisoner's Dilemma chapter* [Ch. 1].

Educational philosopher Alfie Kohn makes the case against extrinsic motivation in schools, at length in his insightful book *Punished by Rewards* [Kohn 1993]. He talks about how the extrinsic mechanisms of grading, report cards, incessant testing, and school choice undermine the intrinsic motivation of love of learning and love of subject matter that truly educate students.

Love is a better master than duty.

Competition and extrinsic motivation

Advocates of competition claim that one of its biggest advantages is that "competition motivates people". But that's a half-true cultural myth.

What kind of motivation can competition provide? Competition can only provide extrinsic motivation, not intrinsic motivation.

Pure competition is a zero-sum game, like a contest where there are no prizes, just the "bragging rights" of being declared the winner. Less pure situations motivate contestants by a combination of these bragging rights, and the motivation of obtaining a desirable prize. In both cases, competition motivates by sparking desires that are completely unrelated to enjoyment or worth of the activity in itself.

The artificial scarcity of "winners" created by contest rules guarantees only a few contestants will have the desire for winning fulfilled, leading to inevitable disappointment in the majority. That disappointment can turn into negative reinforcement of the activity that the prize was supposed to encourage, for the majority. A prize for "good behavior" can thus have an undesirable effect.

People primarily motivated by the social status of bragging rights are rarely the best, most creative, or most qualified. They get tempted to undermine others in their quest for status. Again, also, the loss in social status suffered by the majority branded as "losers" acts as negative reinforcement for the activity.

Competition doesn't motivate all people equally. Competition works best with people who have *competitive personalities* (or so-called "Type A" personalities) which have their good and bad sides: drive and determination, yes, but also aggression and hostility. Competitive personalities tend to be more associated with men rather than women, resulting in discouraging a significant number of women from participating in competitive events. Blanket assertions that people will be motivated by competition tend to disenfranchise those who don't fit the competitive personality profile.

Alfie Kohn makes an eloquent argument against competition in education at length in one of his subsequent books, *No Contest: The Case Against Competition* [Kohn 1986]. Initiatives such as No Child Left Behind and Common Core seek to base all education on extrinsic motivation. A standardized curriculum tells you what to learn at every step, leaving no room for individual interests or intrinsic motivation to influence the agenda. Students are supposed to be externally motivated by getting high scores on standardized tests, not by love of the subject matter. This is the *factory model of education*, which treats students as products of a manufacturing industry.

There are alternatives educational philosophies based on intrinsic motivation, such as Montessori schools [Montessori 1969], Summerhill [Neill 1960], project-based learning, Constructionism [Papert 1993], and High Tech High [Whitely 2015]). The debate in educational philosophy is explored further in our chapter, *Education for Makerism* [Ch. 28]. Kohn's books are required reading for those who want to explore in greater depth our advocacy of cooperation over competition.

Intrinsic vs. extrinsic motivation and the Prisoner's Dilemma

The issue of *intrinsic versus extrinsic motivation* is intimately related to the issue of *cooperation versus competition* that we discussed earlier in this book. There's a tradeoff between intrinsic and extrinsic, just like there's a tradeoff between cooperation and competition.

The two tradeoffs are connected. If people are more intrinsically motivated, they will be more cooperative, and vice versa. If people are more extrinsically motivated, they will be more competitive, and vice versa.

Situations that encourage cooperation will work better at intrinsically motivating people. When we're surrounded by cooperative people, we feel more socially secure. We're more comfortable expressing ourselves, and feel like we can act on our true (intrinsic) motivations without being judged. In competitive situations, don't be surprised if competitors' motivation focuses more on the extrinsic rewards than the intrinsic pleasure of the activity.

Everything we've said about Prisoner's Dilemma situations also applies here. Just as scarcity promotes defection in the Prisoner's Dilemma, scarcity increases the relative power of extrinsic motivation over intrinsic motivation.

Let's say I have two job offers: in one, I like the work better; in the other, it pays me more money. Which will I take? If I'm poor, I'll probably go for the one that pays better. Sure, I'd like to have work that suits me—but first, I gotta pay the bills. If I'm rich, I'm much more likely to value how I'm spending my time, and the salary difference may not make as much difference to my lifestyle. On my goal stack from Maslow's hierarchy, the better-paying job satisfies lower-level physiological and safety needs. The more fulfilling one satisfies my needs for self-esteem and self-actualization.

Conversely, abundance promotes cooperation in the Prisoner's Dilemma, and favors intrinsic motivation. We've talked about how incremental defection in the iterated Prisoner's Dilemma leads to destructive arms races. We can see the same phenomenon in arms races between participants in a competitive contest.

While defection in the Prisoner's Dilemma might be a good short-term strategy, we've seen that cooperation is a better long-term strategy. Similarly, extrinsic motivation can work in the short-term, but intrinsic motivation is better in the long term.

Many situations have a combination of internal and external motivation. Whether someone decides to take action or not is determined by whether a combination of motivations from all sources exceeds their personal threshold for taking action (which might be different for different people). We can express that as the equations,

Total Motivation = Intrinsic + Extrinsic

Total Motivation > Activation Threshold → Action

Everybody understands that somebody has to pick up the garbage. Few people intrinsically like picking up garbage, so it makes sense to introduce some kind of extrinsic incentive for people to do it. Many sanitation workers take pride in contributing to society by doing a necessary job, and the social relations they develop with their neighbors, so in this case the extrinsic incentive isn't necessarily destructive of intrinsic motivation.

You can win fabulous prizes!

Certainly, in some situations such as competitive games, people do seem to get motivated as they become wrapped up in the activity of trying to win the game. Gamers get a shot of adrenaline as they anticipate the thrill of winning. "It's almost like a drug", they say. But, like many drugs, the adrenaline hangover is a bummer. And there's the risk of addiction.

Many educational games try to use students' enthusiasm for games to get them to sit still for learning something. This isn't a bad strategy—in small doses. They turn hated quizzes into guessing games, where correct answers are rewarded by points, animations, increased gameplay time, or places on leader boards. Much better, exploration or simulation games can expose students to situations they might never be interested in on their own.

What that accomplishes is to give the students a taste of the subject matter, in the hopes that they will discover a latent interest. The short-term attraction of game mechanics might get them "over the hump" of reluctance to try new things. But for the lessons to go beyond the game itself, or to persist when the game is inevitably over, the initial exposure generated by external motivation has to be quickly translated into intrinsic motivation. If we rely exclusively on extrinsic motivation for too long, the positive effects of external motivation will have worn off, but it'll be too late for intrinsic motivation to save the day.

In *gamification*, which artificially introduces competition in education and the workplace, people may feel obligated to participate. Those who don't have competitive personalities will actually be *demotivated* by artificially competitive situations. They sense, not incorrectly, that situations that necessarily have few winners and many losers can be a sucker bet.

In the past few years, there's been a fad for contests, "challenges," "grand prizes," etc. in scientific and engineering fields. We have no objection if it's only good, clean fun between consenting adults. And if nobody's job is on the line, and nobody's self-worth is wrapped up in their success in the contest. But on the whole, we think this fad has been detrimental to science.

Contests encourage competitive attitudes and secrecy between contestants. They focus people on incremental progress in very specialized areas, for one-shot tests. Science and other creative fields need exactly the opposite—collaboration between researchers, openness, a diversity of approaches, "out of the box" and long-term thinking. These fields need the freedom to choose what problem to work on, rather than have it dictated by the arbitrary rules of the contest.

It is crucial to contests that they exploit a cognitive bias, the gambling mentality. They want people to imagine that they've already won the prize, and imagine how great it would be. But they know that people have a systematic tendency to overestimate their chances of success. Happy stories of the winners are trumpeted, but the vast majority of losers merely get their time and money wasted. And there are tragic stories of lottery winners who find the sudden influx of wealth more a curse than a blessing.

Contests are said to encourage risk-taking, but there's a difference between risk-taking and gambling—the risk-taker knows the odds, whereas the gambler doesn't bother to figure them out. The gambler is merely mesmerized by the thought of winning.

A few years ago the US Defense Advanced Research Projects Agency (DARPA), which has had a glorious history in the 1970s and 80s of funding innovative work in artificial intelligence, became enamored of contests. Most notably, a contest for a self-driving vehicle, did achieve some success [DARPA 2005].

DARPA crowed about how little they spent to achieve that result, conveniently not counting the unpaid efforts of the unsuccessful and almost-successful contestants. It browbeat researchers who it was funding into participating, turning off many creative people who refused to "gamble with the rent money." We believe it set the field back by years, primarily because of the opportunity cost of distracting researchers from meaningful scientific goals not encompassed by the contest. The most harmful aspect of research-by-contest is that it may convince funders that contests are a substitute for unconditionally funding research. As researchers, we are personally bombarded with countless invitations to participate in contests. We decline them all.

The bankruptcy of incentive

Traditional Capitalism is based on the idea of *economic incentive*. Capitalism works by providing economic incentive for people to do activities that can increase economic output. It views people as merely passive followers of incentive, *Homo Economicus*.

This fails to take into account people's intrinsic motivation, what they want to do or like to do. This is what Marx meant when he said that workers are "alienated" from their jobs.

As we've seen, to some extent, intrinsic and extrinsic motivation inhibit each other. If you want to make sure that people are motivated by incentive, you have to find some way of reducing or sidelining internal motivation. This is one reason why Capitalism can become an instrument of oppression—the whole system is designed to get you to do things you don't want to do.

For traditional Capitalism, one way it does this is to maintain scarcity. Sometimes it's the reality of scarcity, sometimes just the perception of it is enough. If you don't have enough to eat, anything that looks like it might help you get something to eat is attractive, regardless of whether you like to do that thing, or not. Competition pits people against one another, and the external incentive of getting a leg up on your competitor makes you ignore your own needs and desires. As we discussed before, scarcity both promotes competition and is promoted by competition, thereby maintaining the primacy of extrinsic incentive.

On the other hand, in situations of abundance, people are more inclined to follow their own desires and preferences, rendering external incentives ineffective. People are more likely to be cooperative in situations of abundance, which makes them more likely to express their own motivation and consider the motivation of others.

Our central argument is that, because of technology, we're moving from situations of scarcity to situations of potential abundance. That means that we now have the opportunity to move from situations mainly governed by extrinsic motivation, to those driven by intrinsic motivation. In the initial stages of Makerism, the extrinsic motivation of mitigating the failures of industrial capitalism will be an important driver. As Makerism takes hold, people's activities will be increasingly determined by their own personal motivation, and external economic incentives will have less and less force.

Livin' the life of Riley?

People sometimes ask the question, "If, in the future, nobody has to have a job, what will people do all day? Won't they get bored?". That question takes for granted the idea that the only way to prevent boredom is to participate in an activity which is being paid by somebody else.

Because many people have spent their work lives acting almost entirely from extrinsic motivation, in situations that lead them little choice for personal expression, it's hard for them to see how self-direction can lead to a happily productive, active life. But we're confident that the absence of economic coercion will allow people to discover their own interests, talents, and abilities.

There's a big difference between gardening as a hobby and a pleasure, and being wholly dependent on subsistence agriculture in order to eat. There's a big difference between physical effort in athletics, and manual labor necessary to keep yourself alive. The difference between physical effort in rock climbing, and manual labor in breaking rocks in a prison chain gang. Both kinds can be done strenuously, but in the first case under your own volition for goals that you, yourself have taken on. In the second case, that choice is forced upon you, upon pain of death. It's hard to enjoy it in that case.

We can separate the question of what's necessary to meet a person's economic needs from what's necessary to enable someone to lead a happy and fulfilling life.

For a lucky few (including us), their jobs provide intellectual stimulation, positive social relations, and a feeling of contributing to society. But it really is only a few. Studies show [McGregor 2013] that only 13% of people like their work. Studies also identify the most important characteristics that make a job fulfilling [80,000 Hours 2016]. Work you're good at, and that helps others. Work that makes you feel "engaged". Many activities, paid or unpaid, meet those criteria, and it's likely that everybody can find some that suit their interests and talents.

The remainder of the criteria are what might be called "good working conditions": adequate pay, fitting in with your family and personal life, and having supportive colleagues. If we don't need a salaried job, if people have more control over their time, personal life, and social relations, than those won't be an issue.

A good point of comparison is what we now call *retirement*—after 30-50 years of working life, many people in industrialized societies can live comfortably on savings, pensions, or government stipends like Social Security. Some people in retirement do feel more isolated and unhappy, especially because of the social stigma now associated with "obsolescence". We hope that that will change as more and more people become less reliant on jobs.

But there are many retirees that do find personal fulfillment in retirement, through a variety of hobbies, "second careers", volunteer work, more time for family relations, etc. They feel relief in no longer having to "work for the man". Studies on retiree satisfaction have recently shown a decline, perhaps due to recent decline of company retirement benefits, but the most conservative pin it at least at 40%, still way above our 13% for working people. More optimistic studies show it at over 90% [Holland 2016]. Post-scarcity, we'll be able to "retire" at age 0.

For further reading, André Gorz explores these questions in his book, *The Path to Paradise* [Gorz 1985]. For a quick and fun read, we recommend (minus his technophobia) Bob Black's classic essay, *The Abolition of Work* [Black 1985]. We'll leave you with his slogan: "Workers of the world— *relax!*".

Part 3
Can we get along economically?

Chapter 11
The productivity of dead people

Why is the income of an American middle-class professional hundreds of times greater than that of some rice farmer in Africa?

Capitalists are fond of extolling the virtues of "hard work". If you work hard, the capitalist system will reward you. If you're paid by the hour and work more hours, you will make more money. But who works "harder", the American or the African? Surely, the African does. They work longer hours, and the work itself is more physically taxing and perhaps more mentally taxing as well.

Likely the American has a college degree and the African hasn't, but we'd be unwilling to say the American is necessarily more intelligent or deserving than the African. We can't even say for sure that the American has spent more hours in education for their profession than the African, though the African didn't spend much time in a formal classroom. The African had informal education from mentors and careful observation of local conditions, and may have spent an equivalent time doing it.

The sorts of things they know are different. The African would be utterly useless in the American's profession, but put in the situation of farming in Africa, the American would probably starve to death.

Of course, the American professional's labor contributes to technology that is of much higher economic value in the first world than the small amount of rice the farmer produces. The big difference is that the American is embedded in a technological society, and the African rice farmer is not.

The core of technology is clever, useful ideas that have been used to design tools. The use of those tools produces economic value. So one could say that the ideas that these tools embody can go on producing economic value even after the people who invented them are long gone.

Ideas also amplify the efforts of people, thus increasing their economic value. A person can produce more per hour with better technology. Often, the best technical ideas don't increase individual productivity so much as they permit groups of people to accomplish more by working together than they otherwise would, even taking into account that cooperation itself has some costs as well as benefits.

Your authors have made recognized contributions to advance science and technology. But we're not so smug as to say we deserve all the credit for the productivity our contributions might have produced. After all, the vast majority of that productive technology was there before we were born. We just added something to it. Since we're American, we get the benefit both of what we did and what the people before us did. The African, by contrast, didn't start out in a technologically sophisticated society. His world is dominated by scarcity, and scarcity results in resource-draining competition, and poor utilization of mental and physical effort.

Now you know why the American makes more money than the African rice farmer. Basically, the American gets to reap the productivity of dead people. The African doesn't.

Inequality and Incentive

A traditional Capitalist might argue that income inequality is necessary to achieve productivity. If we tried to even out income, then there's a danger that productivity would plummet without the "incentive" for individuals and corporations to obtain more money. But let's examine that assumption.

First of all, incentive is not linear in money. A dollar is worth a lot more to a poor person than to a rich person. Two dollars don't provide twice as much incentive to a person as one dollar.

So we should pose the question: How good is income inequality at proving incentive that does good for society in general? Is money doing its job?

To the extent that the income inequality is due to only some people reaping the productivity of dead people, the answer is: Not at all. Cesar Hidalgo [Hidalgo 2015] introduces the more general notion of *Topocracy*, reaping rewards based on your position in a network of interacting people, rather than on your own intrinsic productivity. It's Topocracy that accounts for much of the income differential between the first world and third world.

In the chapter *Intrinsic and Extrinsic Motivation* [Ch. 10], we critique the notion of "incentive" itself as a motivating force. Incentive can only provide an extrinsic motivation, which we argue is not as effective as intrinsic motivation in the long-term.

The Fundamental Theorem of Capitalism

Laissez-faire capitalism has an underlying assumption that economic incentive promotes the general welfare. We call this

The Fundamental Theorem of Capitalism (FToC):

If everybody acts according to the economic incentives that Capitalist society provides, the result will be best for society in general.

Of course, it doesn't mean best for every single person at every moment; just that it would be better than if another overall policy, such as Socialism or Communism, were adopted.

This is, put in different terms, like what in economics is called the *Fundamental Theorem of Welfare Economics* (although it is not about the social support programs called "welfare").

The Prisoner's Dilemma shows that there are situations where everybody acts narrowly in their own self-interest, and the result is provably worse for everybody. This is a mathematical result. It is not a political position, and it is not debatable. Furthermore, as we show in the chapter *Jailbreaking the Prisoner's Dilemma* [Ch. 1], these situations are not rare. They are what cause the worst excesses of industrial capitalism.

It's even worse when the financial incentives encourage someone to do things that cause negative impacts for society in general. The boss getting overpaid relative to workers below them may be unfortunate, but a weapons manufacturer who makes money off of blowing up thousands of people is a tragedy of the first order. In theory, lawsuits could be brought by those harmed by externalities, but the legal system is itself a scam with perverse incentives, as we explain in the chapters *Justice* [Ch. 30] and *The world's best business model* [Ch. 12].

Advocates of laissez-faire Capitalism, who are the political group who hold most strongly to the FToC, seem to be oblivious to the challenge that the Prisoner's Dilemma poses to the FToC. We don't fault the original thinkers, like the Austrian economists, or even Ayn Rand, who all worked long before the Prisoner's Dilemma was popularized with Axelrod's 1984 book. Ironically, Libertarians love to quote the Tragedy of the Commons, which is a corollary of the Prisoner's Dilemma. But, inexcusably, we can find no contemporary discussion in the Libertarian or conservative literature about the general case.

We're all heirs to the fortune of technology

So what *should* happen with the productivity of dead people?

A rich person can leave an inheritance to their children, because it is one of the benefits to rich people to feel like their children are being taken care of. But should this go on forever? What's the point of providing additional incentive to dead people?

An inheritance may or may not be a blessing for children who receive it. In many cases, it provides a disincentive for being productive. The rich are not necessarily happier than those of modest middle-class means that have basic needs met. Some enlightened rich people, like Bill Gates and Warren Buffett, are wary of the corrupting influence of inheritance and have pledged to give away most of their wealth to charity while ensuring a comfortable but not outrageous lifestyle for their family.

Unless it is used wisely, excessive riches may simply fuel needless consumption and have a negative impact upon society. A conventional solution is inheritance taxes, and that works to some extent. Perhaps the US should have a much higher inheritance tax, as some other countries do. Some of the wealth makes its way back to activities that benefit society as a whole, such as rich people funding the next generation of innovative startups. But a lot of it doesn't.

Piketty [Piketty 2014] documents the economic mechanisms by which accumulation of capital fuels more accumulation of capital. When people whose labor is vastly amplified by technical ideas die, benefits spread to their relatives, their employees and colleagues. To a much lesser extent, their customers and the citizens of their society may derive some benefit. It's just that they don't reach far enough and fast enough to reach that poor African.

Some of the creation, exploitation and harvesting of technological ideas is best thought of as done by companies or organizations, rather than individuals. And large companies (IBM, GM, etc.) at least potentially, can last far longer than a human. There's no such thing as an inheritance tax on companies, so companies can continue to amass inherited wealth across generations.

The good news is that you can't keep good ideas bottled up forever. Word leaks out; ideas spread virally; technology is re-invented elsewhere. We are already seeing the fact that the reach of the Internet into even the poorest areas of the third world is spreading innovation faster than it can be restricted by such censoring mechanisms such as "great firewalls" or "intellectual property". The chapter on *Makerism* [Ch. 15] shows how the spread of ideas will result in the spread of material wealth.

Since the vast majority of the technological infrastructure was here before any of us were born, we can consider it the common heritage of humanity. We're all its heirs. If you've been born, you deserve a share. (But you don't get two shares from being "born again"). It is an endowment that produces a dividend, and that dividend is what should fund a basic income for everybody.

Your check's in the mail

The obvious mechanism would be to establish some kind of minimum income, guaranteed income, or "negative income tax". Even such ardent supporters of capitalism such as libertarians Milton Friedman [Friedman 1962] and Freidrich Hayek [Hayek 1994] are on board with this idea.

Anthony Atkinson wrote a book called *Inequality: What Can Be Done?* [Atkinson 2015]. You can see a detailed review by Thomas Piketty here

[Piketty 2015]. The book discusses various forms of wealth transfer: Progressive taxes, inheritance tax, property transfer tax, etc.

We note that this idea can be thought of as just a stepping stone to a more radical solution, as we discuss in *Makerism* [Ch. 15]. Makerism, the alternative that we advocate as the replacement for Capitalism, reduces or eliminates reliance on Capitalist notions of jobs and money, likely obviating a need for such income. But since it'll be a while before full Makerism kicks in, the interim solution is worth exploring.

Alaska has a small unconditional distribution to all of its citizens to share the wealth generated by its natural resources. Switzerland just voted it down, but the fact that the proposal got that far is encouraging. Many industrialized countries have some partial version of it, such as France's *Revenu de Solidarité Active*. You could even argue that the US Social Security SSI/SSD is a form of it for senior citizens and the disabled.

There is also a strong feminist argument for a guaranteed income. Work by women in child care, care of other family members, food, cleaning and other household maintenance, goes unpaid. Work in the public workforce by men is compensated, and men often shirk their fair share of household duties. The result is that many women have no financial independence, distorting personal relationships in the family. A guaranteed income would at least assure that everyone had some personal resources [Shulevitz 2016].

Eventually (and perhaps this is possible already), a universal minimum income will provide enough so that people will never feel like they are *forced* to work. They will perform paid work when they need or want extra money, or when they feel that the work is a positive contribution to their lives and helping others.

Some might find that our utopian vision of an abundance economy supported by artificial intelligence software and the hardware maker movement to be a far-off pipe dream. But the reality is, if we assess the current situation and project future trends, it's really within reach.

First, let's look at how productive the world really is at the moment. According to the CIA World Factbook [CIA 2013], the Gross Domestic Product (GDP) of Earth in 2013 is approximately $77 trillion. If we stopped spending money on unproductive activities such as war ($13.6 trillion

[Schippa 2016]), and invested properly in things like education, that number would be likely much, much higher. Besides the destructiveness of war, military spending displaces investment in education, infrastructure, etc. which would likely yield returns in productivity.

That's really a pretty good number when you think about it. Since there are about 7 billion people in the world, if we divide that number down by the population, that's on the order of $11,000 per person (around $13K if you deduct the military spending, above). Since that 7 billion figure also includes children, we can think about that number as being $44,000 for a family of four. It should certainly be enough to provide a decent lifestyle. The US median household income in 2013 was $51,939 [Wikipedia 2016a].

By any standard, the conclusion is inevitable that the total productive output of the world at the present time is sufficient to support a reasonable income for everybody.

Can money buy happiness?

A question to ask is what is the relation between between income and happiness? After all, happiness is what we are trying to achieve, not just maximizing income. Studies show that indeed, the more money that people have, the happier they are. But only up to a point.

And that point is surprisingly low. Poor people are indeed unhappy, and within a certain range, happiness is linear in income. But after a while, it plateaus, and additional money doesn't necessarily buy additional happiness.

One study pegs that number at household income of $50,000 a year [Sanburn 2012]; another $75,000 a year [Kahneman 2010]. Within spitting distance of our $44,000 share, and certainly what we could get if we do things like eliminate war. Even at the high end, these are still modest middle-class incomes by American standards. Of course, you can argue about how these things are measured, factoring in differences in cost of living in various places. But the overall thing to note is that these numbers aren't very far from the quantification of the average productivity.

Poverty is too expensive

Actually, if we think about it, we can't afford *not to* provide a basic income for everybody.

If we don't, we inevitably have poverty. You might think that poverty doesn't cost society much because poor people don't get paid much. But you'd be wrong. Poverty causes *externalities* that cost society vastly more than anything we could hope to save with low salaries.

Where will poor people live? Why, in slums, of course, where housing is cheap. Nobody's figured out how to have poor people without having slums.

Once you have slums, you have higher rates of crime. Maybe it would be possible to have slums without having a higher crime rate, but conservatives can't tell us how to do it reliably. Similarly, higher rates of harmful drug use, higher rates of medical problems because of the lack of preventive care, poorer education, and a host of other problems. They can't be divorced from poverty itself, because poverty is the cause.

So we have to chalk up to poverty all the costs it creates: the costs of policing, the cost of crime, the cost of the "justice" system, medical costs absorbed by society, etc. etc. It costs way more to keep someone in prison than to educate them, or to get them into a productive job.

Internationally, you could argue that poverty is such a root cause of war that we also need to chalk up all the costs of the military and waging war, on poverty's tab. There's also the opportunity cost of losing productivity from people whose time we waste with dealing with all the problems of poverty. Worst of all is the human cost in needless suffering.

Why do we tolerate it? We shouldn't. Poverty is so expensive, we can't afford it.

The surprising fact is: we can get an enormous amount of help solving the worlds' problems from dead people. We just have to unleash the ghosts.

Chapter 12
The world's best business model

Get ready—as a special bonus, we're going to tell you about a seldom discussed business model which, if you can pull it off, will yield an entrepreneur untold riches. It's already been proven by a wide variety of successful companies, including Google, Facebook, banks and financial institutions, advertisers, and many others. Some might consider it slightly unethical. But, if you can put that aside, we can pretty much assure you it will work and that there is no practical way somebody can come after you for it. Go for it. You can thank us later.

Here's how it works. First of all, remember the world's very first computer crime? It was a programmer at a bank who was trying to compute the interest on bank accounts. There was roundoff error of a fraction of a cent. He suddenly realized that he could write the program to transfer that fraction of a cent from every customer's account to his own account. The customer would never miss a fraction of a cent. The bank's balance sheet would add up at the end of the day. There would never be any money missing that somebody would try to track down. Voilà! Instant riches! They only got the guy when he made the mistake a lot of criminals do: he started spending lavishly and the tax guys started to wonder how he was able to do it on such a modest salary. We call this model *aggregation of microtheft*.

When you speak to executives of large corporations, they are usually very proud of their "brand equity", the trust that they say consumers place in the recognizability of their brand name and their reputation. But are the companies deserving of such trust?

Actually, yes and no. If you put $10,000 into your bank account, you can certainly trust the bank not to abscond with all the money. But can you trust the bank not to steal $10 from you? Everybody's had the experience of finding an unexpected $10 "service charge" on their account, where the service in question is in fact rendered only to the bank, not to the customer. You can try to call them up and complain, but that has little hope of success, and is guaranteed to cost you at least a half hour or more of your time, worth at least another $10. Congratulations. You've just been micro-mugged.

In general, companies can be trusted not to commit grand larceny, but petty larceny is an integral part of their business model.

Aggregation of microtheft is much easier to pull off if what you're stealing is intangible, like the customers' time on a phone call. For companies like Facebook and Google, you're not the customer: you're the product. Facebook's "product" is the content produced by all their users, who are not being paid for it. It's the "eyeballs" of their unpaid users being sold to advertisers. Google's search engine works so well because it leverages the work done by millions of unpaid webpage authors who create web links that Google can mine for popularity.

Facebook and Google do deserve some credit for being a central place where this kind of aggregation can occur, and for providing user interfaces and algorithms that can take advantage of it. But, by the standard of labor that it takes to produce value, they certainly don't deserve to reap all the rewards. The number of man–hours spent by the people who contribute to Google and Facebook would certainly dwarf those of the corporate employees. It's just that each individual contribution is so small that society has not been able to make any practical mechanism to assure that individual contributors get paid. So each fraction of a cent accumulates to the aggregator. Sound familiar?

Advertising is itself aggregation of microtheft. It is theft of attention. (One could argue that just *looking* at ad-supported media is implicit acceptance of an attention-for-content trade; but like software EULAs, the degree of actual consent is debatable. Certainly something like a street billboard implies no consent).

You can understand this by analogy to the crime called *identity theft*. Why is identity theft a crime? Well, identity is all you have as a human being. If you steal somebody's identity, you can steal all their money, and they have nothing left. It took a while for society to recognize identity theft as a crime and pass laws against it.

Likewise attention. Like identity, from another point of view, attention is all you have. If somebody steals your attention, you can't pay any attention to anything else, and you have nothing left. It's just that the amount of attention stolen by advertising is so small that most people can ignore it. *Attention theft* is not (yet) a crime in our society. Maybe it will be.

Advertising can also be considered *microfraud*. It's not as bad as outright (macro-)fraud. We do have laws that prohibit patently false statements in advertising (but these laws are not often enforced) and control what advertising can say (but not strongly enough). It's "micro" in the sense that advertisers always seek to exaggerate, mislead, and insinuate, but they're usually careful not to make their misstatements so blatant as to trigger the laws against fraud.

Advertising causes us to make purchasing decisions that aren't really in our best interests and sets us up for disappointment. As such, it has a negative impact on the value we receive for the products and services we buy, so it has a net negative impact on the economy.

In *Jailbreaking* [Ch. 1], we used the example of advertising to illustrate how advertisers use vendors' fear of competition to cajole them into spending money on ads, even when it manifestly doesn't increase sales or market share. This isn't a small thing. Direct spending on advertising in the US was around $200 billion [Statista 2015].

Ads do serve two purposes. The *cooperative* purpose is to match buyers and sellers, and in the event of a successful match and the buyer appreciating the product, they are positive. But now, we have resources like consumer forums, third-party review sites, collaborative filtering, comparison shopping agents, etc. that can serve the function of decision support for purchases, and are less biased.

The *competitive* purpose is to trick the consumer into spending more than their legitimate needs and desires would warrant, and often result in dissatisfaction with the purchase and unnecessary depletion of the consumer's resources. Ad agencies spend untold effort on just this trickery. In these cases, the entire cost of the product should be counted as waste caused by advertising, because it shouldn't have been produced and sold to that person in the first place. It also distorts the *price signal* to producers: sales goosed by advertising result in overproduction of goods, which can only be sold by more aggressive advertising, in a vicious cycle.

The consumer's time watching the ad, and opportunity costs for both producers and consumers should also appear in the negative column. It's just that nobody can be held responsible for the microfraud of advertising, so there's no easy way to recover from it.

One traditional mechanism that is supposed to address the problem is class-action suits—in theory a group of victims of microtheft or microfraud could band together and sue the aggregator. In practice, class-action suits are a joke.

Except in isolated cases, they have been completely ineffective in dealing, say, with environmental problems. Members of victorious classes in class-action suits against faulty products generally receive only token compensation. As in other legal matters (see *Justice* [Ch. 30]), the only true victors are the lawyers, who get outrageous fees. You're the victim of microtheft again, only this time, it's at the hands of the lawyers.

We think aggregation of microtheft is one of the central bugs in classic Capitalism. Like we saw in the Prisoner's Dilemma [Ch. 1], it's yet another instance of how a mechanism that seems locally plausible (aggregation) can result in a globally poor outcome. It's hard to imagine what kind of specific laws or regulations we might pass that might effectively outlaw aggregation of microtheft. Solutions will have to await the larger transformation of Makerism that we advocate.

Some microtheft can be dealt with by creating alternatives to rapacious aggregators. Maybe credit unions and cooperative banks won't be motivated to charge bogus service charges. Maybe peer-to-peer alternatives could replace Google and Facebook. Maybe consumers will wise up and make purchases based on third-party review sites, word-of-mouth, and recommendation agent software, thereby disintermediating advertising. We can at least dream, can't we?

Anyway, now you know. If you start a successful company based on what we told you, remember: we want a cut of the profits.

Chapter 13
Slapped by the Invisible Hand

You go to Home Depot looking for a lamp dimmer. Hmm, with 40K different items, you're pretty sure dimmers are here somewhere. A list of departments would help, but they only have signs on aisles and you can't see all of them at once. Maybe the electrical department? Where's that?

No big box stores have a map with an index. A department store mid-morning, mid-week, is a ghost-town, lacking readily available staff. The little stores are much more customer friendly, but the big ones are taking over. Since all the big stores claim customers are their focus, why don't they have maps? If Capitalism is so efficient, why does it waste so much of consumers' most valuable resource—*time*?

The myths of Capitalism

Capitalism is the dominant economic system of our times. It values private ownership, (especially of the means of production) trade, and a free market. In the USA, Capitalism rivals any religion for the intensity that people believe in it, perhaps second only to the Constitution and voting for its popularity. Is this popularity deserved?

Here we explore the *myths* of Capitalism: widely held beliefs that are, at best, only half-true in the real world. The worst problem with these myths is that they impede the search for fixes or alternative solutions.

Myth 0. In Capitalism producers cooperate with customers.

The beauty of ideal Capitalism is that the mechanics of the marketplace can act as an *algorithm* for meeting needs of both the producer and consumer. There's no need to negotiate each case individually.

Adam Smith, the inventor of modern Capitalism, called this the *Invisible Hand* [Smith 1776]. The problem is: the store also *competes* with the customer, in a Prisoner's Dilemma-like fashion. Smith, writing in 1776 (!), didn't know about the Prisoner's Dilemma.

The common wisdom on why big stores make it difficult to find a product, is that they want you to wander around the store and buy something you didn't think you wanted. Presumably, somebody's done a study, and determined that impulse purchases happen often enough that the store makes more money than they lose from frustrated customers. That's the *Temptation* that causes the store to *defect*. **The customer comes last** is a theme throughout retail. (Ever buy an airline or concert ticket from a website?)

The store wins. The customer loses. The customer suffers an *externality* of wasted time and buyer's remorse on impulse purchases, but the store doesn't care. We call this kind of defection **Myth 0** because it's the root of the problem. The other myths will follow from this. Why can't you find the lamp dimmer? You've just been slapped by the Invisible Hand.

Myth 2. Capitalism is efficient.

What's so efficient about losing a sale due to no store map, or to bad web programming? Could making it easy to buy their products give them 1% more business? Easily. 1% of Home Depot's $83B annual revenue is $830M, certainly more than the implementation cost of a map. Leaving money on the table is so common in business that it points to systemic problems, not individual stupidity.

It's difficult to get average numbers on this, but roughly 25% of the purchase price of a retail product goes to the manufacturer. The rest goes to transportation, warehousing, retail, marketing (including all the junk mail you have to pay to get hauled away to ruin the environment), corrupt politicians, etc.

With so many fingers in the till, we wouldn't call Capitalism "cost-effective". But compared to what? Compared to Makerism, coming to a chapter near you.

Myth 3. Build a better mousetrap and the world will beat a path to your door.

The primary business motivation is not to produce the best possible value to their customers. It's to maximize profits.

Innovation is useful to gain *competitive advantage* over other companies or extract more dollars from a customer. Improving customer experience is irrelevant. Dominate players hate innovation because change threatens their position. Small innovative companies are something to be bought to kill, or in the best case, bought to help a company keep up with other large players in their field.

Competitive advantage only requires a *minimal* amount of innovation—just enough to make your product better than the other guy's. Because major innovation is risky, significant innovation is slowed by Capitalism.

Startups and Venture Capital are supposed to be the engine of innovation. But the startup process selects for copycat companies that are only incrementally different than already successful business models. Now, do we need another *SnapChat, Twitter,* or *Share Irrelevant Minutia To Strangers* company? No, but we'll get one next month anyway.

Our society also chronically underestimates the *urgency* of innovation. If you believe the clock is ticking on climate change, war, and your own health; then, our slow pace of innovation is not merely inefficient. It's fatal.

If the Web is so great, how come programming languages and environments are still so hard to use? Why don't we have Personal Rapid Transit (See *Transportation* [Ch. 29])? How come innovation hasn't reduced the percentage of income people spend on food, housing, transportation, education or health care (*all* the essential, big ticket items)? 13% of Americans (in the richest country on Earth) live in food-insecure households [Feeding America 2015]. Sorry, even the "most innovative country in the world" isn't innovative enough.

Myth 4. Intellectual Property promotes innovation.

Developing new ideas is crucial for advancing civilization. When Capitalism started to become widespread, it became clear that the drivers of economic productivity weren't just labor, land, and raw materials, but also invention and innovation (see *The productivity of dead people* [Ch. 11]).

At that time, Capitalism seemed to be doing a pretty good job of building an economy based on material goods. So, using the powerful heuristic of problem solving by analogy, people thought: Can we make an analogy between inventions and material goods, so that people can buy and sell them, and inventors can earn a living?

Only problem is, the analogy breaks down. While material goods are tangible and can only be in one place at one time, ideas are intangible, and can be infinitely copied at zero cost. Ease of copying ideas is actually a strength—it's what *enables all human progress*. In education, after all, ideas are "copied" from a teacher to a class of students.

But you can only "sell" exclusivity of access. So the oxymoron of so-called "intellectual property" was introduced. Patents and copyrights are the legal mechanisms. This seemed like a good idea at the time, but now it's just welfare for lawyers.

Intellectual property is defended as being for the benefit of creative inventors. But very, very little winds up in the hands of inventors. Only 2% of patents make any money at all. Only a very small percentage of that number makes any money *for the inventor*, after you subtract the outrageous legal expenses and exclude patents assigned to an employer by an employee who has no rights and often receives no benefit. Our guess is that if an honest reckoning were done, the "return on investment" for inventors on patents would probably be less than minimum wage.

On the other hand, it's a fantastic business model for patent lawyers, who charge $500/hour or more. Our university spends an average of $12K to file a patent, not including any litigation needed to enforce the patent. Like arms merchants, lawyers rake it in, regardless of who wins or loses the war. Patents do not give the holder the right to make money off the invention; they only provide the right to sue others who do. To sue, you got to... pay a lawyer. And if somebody sues you, you've got to... pay a lawyer. A typical patent infringement suit costs $500K or more.

There's yet another inefficiency here: if there is some physical process that really is better than all the rest that are known, then if one company locks that up, they win big.

Say the process makes cost to manufacture 10 times cheaper. They don't sell the product for 10 times less, but just slightly under their nearest competitor, denying society most of the benefit of the breakthrough.

Now if another competitor comes along with an entirely different process that is also 10 times cheaper, what happens? Well, the "invisible hand" says that both companies drop their prices to just above costs and society finally wins from the first big invention, late though it may be.

But other outcomes under Capitalism are likely such as: The bigger of the two companies kills the new company via tying them up in legal work, predatory pricing, buying and burying the new invention, collusion with the other company to maintain high prices, or even less desirable means.

Imagine though that the first new invention is made available for others to build on. We could get more rapid advancement.

Myth 7. The "free market" actually exists.

The reason there is no free market (never has and never will be) is because no one wants one. Economists will tell you they do, but then ask you how to get a special deal on a new car (thereby bypassing the "fair market value"). Libertarians will tell you they do, but they really mean "no government meddling".

So what do you do when Manufacturer A pays the distributor to not carry Manufacturer B's product? Or big oil destroys the environment? If the answer is "Nothing", then you get US Capitalism, not a "free market".

Myth 8. You get what you pay for.

In order to get what you pay for, you at least have to know how much you're paying. US Capitalism makes this hard through the time-honored process of *Bait and Switch*.

Most people think they won't buy something without knowing its price. Rarely is the true price available to a purchaser until after the sale. First, in the USA at least, tax is not included in the listed price. Why? Because the seller wants you to believe the price is lower than it actually is to encourage the sale.

Why do we have "tipping" in restaurants? Studies have shown that tips do not significantly affect quality of service [Lynn 2001]. What tipping accomplishes is to misrepresent the true cost of the meal, and to shift the guilt for underpaying workers from the restaurant owner to the customer.

This kind of dishonesty ought to be fixed by government. But government *also* wants you to buy the product, so they can get the sales tax. Retail sales are a collusion between the seller and the government against the buyer. Due to special interest politics, the concentrated power of a seller is much more significant than the larger, but diverse, power of consumers. Hence laws are written for sellers, not buyers.

In any complex product, say a washing machine, there are a whole bunch of other fees that get added on after the customer has decided to buy. Car buying, for example, is notoriously riddled with such dishonesty from the seller. Figuring out how to lower the price you think you are paying and raise the price you actually pay is called "sales creativity".

Myth 9. Employees are compensated for what they're worth.

The average pay of the CEO's of the 350 largest companies in the USA in 2015 was $16M. That's 300 times their worker's salaries at $53K. OK, the CEOs are smart, but 300 times smarter?

Let's examine another kind of employee, one that gets paid $0. Hamid Ekba and Bonnie Nardi have coined a new term *heteromation*, [Ekba 2014] to refer to the now common practice of getting consumers to do much of the work of the product they buy When you wade through the voice menus of customer support, you are doing the work of a receptionist. When you "assemble" the pieces of Ikea furniture, you do the work of a factory worker. When you pump your own gas, you do the work of a gas station attendant.

Such activities now permeate modern Capitalism. They *may* lower the cost you pay, enabling you to trade your time for money. Or they may simply raise the profit of the "manufacturer" by reducing their labor costs. Capitalism suffers from not compensating workers fairly, including consumer-workers. This trend is not sustainable.

Myth 10. Capitalism allocates labor skills efficiently.

When an economic good is "ripe" for development, often numerous companies jump in. Now, often there's room for a few market entrants, but many dollars are wasted in the duplication of discovery and invention that could better be spent on unique things, *or* the researchers in an area could collaborate, and get to market even faster than the first, with an even better product.

Myth 11. Competition drives prices down to just above production costs.

Adam Smith's *Invisible Hand* of Capitalism would make the prediction above. But no producer wants to sell their wares slightly above what it cost to make them, they want to sell them just barely below their competitor. If they manage to kill off their competitor, that means they can charge much more.

Government tries to ensure a "level playing field", but since government regulators are largely staffed by the industry they regulate, corruption is inevitable. Lawyers of course love "cheating" of any form as they have a near monopoly on "resolving" it. Profits are high (Exhibit A: Inequality) and the playing field is just about as "level" as the Rocky Mountains.

Myth 12. Capital flows to where its most needed.

The highest ROI in American Capitalism is ... *paying bribes to governments to give businesses unfair advantages.* [Bernabe 2015] documents the return on investment of lobbying for American's 200 most politically active companies was 7600%. That is, for every dollar spent, the return was $760.

Nor were these investments small. By investing $5.8B, big businesses received $4.4T or 2/3rds of what individual taxpayers paid the government for 2007 through 2012. (Now you know where your taxes go.) By comparison, the average ROI of a Dow Jones stock in 2010, a legitimate investment open to anyone with the cash, was 11%, or 690 times less. Such corruption leads to large scale economy-wide inefficiencies. We do not need more money in bribes.

Myth 13. War is good for the economy.

The most capitalistic country, the USA, sells the most weapons. (1/3 of global weapons supply.) It sells them to half the countries of the world. [Oakford 2016]. Many of those weapons went to the Middle East.

Yes, you'll hear that trade creates friends. But economic competition can lead to war. We can see how well the trade of those weapons are keeping the peace. Selling to both sides of wars is, of course, a winning economic strategy: what you sell gets blown up by what you sell. You can't saturate the market.

Let's follow the money: Taxpayers to governments to militaries to weapons manufacturers to weapons given to people who use them to blow up the weapons, people, buildings and a bunch of other wealth as well. People get angry at the destruction leading to hate and more war. How is this good for an economy?

Myth 15. Consumers buy what they want.

In order to buy what you want, you have to know what you're buying. Department of Energy to the rescue. The DOE has a program to help consumers choose energy (and water) efficient appliances. So far so good. But their rating mechanisms are questionable. For washing machines it includes energy costs, but as most conservationists know, washing your clothes in cold water is perfectly fine, so the DOE numbers don't help much. The DOE doesn't even bother to rate ovens. Store sales clerks told us "all ovens are the same". But induction cooktops are more efficient than resistance cooktops (according to manufacturer claims) and doesn't the insulation thickness determine how much heat an oven retains? At three big box stores, we found the salespeople hard to find and remarkably unknowledgeable about their products. We estimate that only 10% of showroom floor models for washing machines, dryers and refrigerators have DOE energy usage tags on them.

Myth 16. Capitalism makes people richer, which makes them happier.

The Happy Planet Index (happyplanet.org) measures people's self-reported well-being and their life expectancy per their country's ecological footprint. The USA rates as the worst country in the Americas on this scale, because, although it has a high life expectancy and well-being, it is in 105th place in ecological footprint per capita.

Myth 17. A rising tide lifts all boats.

Economies are enormously complex. Capitalism has a rather elegant way of simplifying: money. Cost, price, interest, stock, and GDP quantifies, and thus makes manageable, a huge range of complexity. But there's a cost to this simplification, which we can fit under the umbrella term of "externalities".

Exhibit A: One car doesn't make much of a difference to the planet. But in 2010, the world's car population exceeded 1 billion [Tencer 2011]. The consequences? In April of 2015, for the first time, the atmosphere maintained 400 parts per million of carbon dioxide for an entire month. Now, it's permanent.

Ultimately, externalities cut the efficiency of the economy: sick people can't work, depleted resources limit manufacturing, and dishonest advertising reduces our ability to make optimal decisions. In the 1980s this was called "trickle down economics" (See the *Ultimatum Game* chapter [Ch. 2]). We won't rehash Piketty's arguments here [Piketty 2014], but in short, inequality is rising a lot faster than the boats.

But boats are rising, nevertheless. That's because global warming is causing sea levels to rise. Three quarters of all large cities are on coasts. Global warming is caused by, you guessed it, economic activity. Capitalism is addicted to oil; even George W. Bush said so. Conflict seems to follow oil (the Middle East, Nigeria, Venezuela). The US sponsors war all over the world, and one might claim Canada's tar sands oil extraction are a war on the planet [Klare 2005].

Myth 20. Capitalism is a good decision-making process.

Trump declared bankruptcy 6 times. He claimed doing so were good business decisions, and maybe they were—for him. He was just "taking advantage of the law of the land". The contractors he stiffed weren't so happy with that decision making process. We shouldn't be, either.

Companies hire CEO's for salaries of millions of dollars on the premise that the right leadership can help them make decisions leading to money. Maybe so, but the question is, how many people are hurt by high CEO pay? Concentrated wealth tends not to allocate the pie with the greatest good for the greatest number.

Since Capitalists make money by exploiting cognitive biases (see the chapter, *The world's best business model* [Ch. 12]), it provides perverse incentives that encourage departures from rationality.

Capitalism, like US Democracy, values short term decisions. But most of our big problems are not short term. They require strategic thinking decades into the future. Capitalism appears to be immune to reason.

Chapter 14
Can Capitalism be saved?

Most economists would say Capitalism is stronger than ever. We're not most economists. What's the problem?

Capitalism's life expectancy

Capitalism is committing suicide. This is not an accident. It is woven into the very fabric of the institution. All capitalisms must sow the seeds of their own demise by the very processes that make Capitalism Capitalism. Marx and Lenin both thought Capitalism was doomed. Their predictions were off by 100 to 200 years, their reasons incorrect, and Capitalism certainly outlasted Communism.

But there's a new threat. Businesses, especially large ones, must ultimately be more profitable than their competitors, or the competitors will buy them, or drive them out of the market.

$$Profit = Revenue - Cost$$

Increasing revenue is harder than cutting cost. The easiest way to cut costs is to cut labor. The easiest way to maintain production and cut labor is to replace people with robots.

This is happening in nearly every industry. In the past, it has been possible for labor to shift to new jobs that machines are incapable of. We believe that automation will soon reach a point where there's little left to shift to. As machines get both more dexterous and smarter, fewer people will be able to compete and thus they become permanently unemployable.

Without employees there will be no paychecks.

Without paychecks there will be no consumers.

Without consumers there will be no customers.

Without customers there will be no businesses.

The very businesses that laid off their workers to increase profits, will have goods and services to sell, but no one able to buy them. As businesses fail, we go into the dark ages, or change the economic system. Either way, Capitalism dies.

You can read a long-winded explanation in [Ford 2009]. Another version is in [Rifkin 2014]. [Santens 2015] predicts the demise of one of the most popular professions: truck drivers, due to automated trucks. The even more popular profession of teacher is under attack from on-line courses like EdX (Harvard and MIT's free online education site). *Lights in the Tunnel* predicts that no profession is immune to the onslaught of automation.

Tweaks to Capitalism

Let's explore some possible transformations of our economy that others have proposed to get us out of this jam.

Stop Automating

The Luddite cry heard as far back as 1811 [Wikipedia 2017b], is the first defense to consider. If a company doesn't automate, someone else will. Consumers will buy the cheaper product and our "humane" non-automating corporation goes out of business, thus having to lay off the very workers it was trying to protect. Simply put, we can't stop automating and maintain todays' Capitalism. Also few would want to give up the wealth-generating benefits of automation.

Retrain Workers

This is the favorite response of politicians. Retraining has worked for hundreds of years. But there are signs that it won't continue to work. First, trends don't go on forever. Second, there are two primary areas that have resisted automation, human physical *dexterity* and *human intelligence*. But this resistance is now breaking down, due to more precise robots and progress in AI.

Income Redistribution

Universal Basic Income (UBI) has been proposed as a way to reduce welfare costs. The money for it would presumably come from taxing the rich, or corporations. It might work. Bernie Sanders was fond of saying, "When the 1% own as much wealth as the bottom 90%, I call *that* income redistribution" [Sanders 2016].

UBI requires companies to pay more taxes. What corporation wants to do that? And the major corporations influence their own tax rate. It is possible that Capitalists reluctantly agree to it if the immediate alternative were civil unrest, as they agreed to the New Deal when the Depression caused massive unemployment.

We think UBI is a feasible shorter-term solution, for the case where a minority of people would be out of work due to automation. However, it is unlikely to work when most people are laid off because the drastic change in tax laws is just not politically feasible in our current economic and government structure. It also won't quell civil unrest due to inequality and the likely corruption and dissatisfaction of citizens with a corrupt government made more probable by the expanded volume of money filtering through government's fingers. In either case, any economic system that adopted UBI would not be Capitalism by definition. We explore the idea of a guaranteed income in greater depth in *The productivity of dead people* [Ch. 11].

Big changes

If tweaks won't do it, perhaps something more drastic will.

Communism

Communism is a socioeconomic order structured upon government ownership of the means of production, absence of social classes, and the ambition, at least in the long term, to get rid of the state. Communism as an ideal of cooperation and equality, has a certain appeal. Unfortunately in practice, it has led to totalitarian regimes that limit freedoms, and are less efficient than Capitalism at allocating resources. China has done amazingly well economically in the last few decades, but this is largely due to the extent that it has abandoned Communism and become more like Capitalism.

Socialism

There are many forms of Socialism, but broadly, they fill the spectrum between Capitalism and Communism. Socialism may have democratic government, some form of individual compensation for work, yet still have the means of production be collectively owned either by the nation or smaller organizations and at least some central planning. The right mix of features in Socialism can improve upon aspects of both Communism and Capitalism. It hasn't proved great for innovation, breaks no new ground in decision making processes, misses the nuances of individual choice, and hasn't replicated itself as widely as Capitalism. If Socialism had solved our major problems, we wouldn't have written this book. None the less, its better ideas deserve attention.

Plutocracy

Dyed in the wool Libertarians would argue that the problems with Capitalism are not the fault of Capitalism per se, but rather our "watered down" implementations of Capitalism due to too much government control. They claim that the "free market", if left to its own devices, would fix what apparently ails Capitalism.

Without controls, the inequalities inherent in Capitalism become extreme. This leads to Plutocracy, a form of oligarchy and defines a society ruled or controlled by the small minority of the wealthiest citizens (Wikipedia). The United States is going through an experiment in the "merits" of Plutocracy right now. We see even poor, yet uninformed, citizens voting for Plutocracy. It tends to undermine Democracy, particularly minority rights.

How do plutocrats get the rest of the citizens to give up their wealth and rights? Fear is the strongest motivator. They convince citizens that there's some bad guys to be afraid of and the only way to protect yourself is to give the plutocrats more power. Its made easier when there's scarcity amongst the population, ironic because the plutocrats will take even more of the citizen's money.

Implausible? Hitler used the fear of Communists to become a dictator. Britain's right wing used the fear of immigrants to defect from the EU and gain power for themselves. The fear candidate in the 2016 US Presidential election beat the hope candidate with a party that is undermining Democracy [Krugman 2016].

No good options?

We covered the most commonly proposed tweaks to Capitalism. Many more have been proposed by others. We don't perceive any as being sufficient, implementable and still qualify as Capitalism. Nor have the "big changes" to other forms of government worked to date. We believe tweaks to them are also unlikely to solve our major problems. Fortunately there's another option: *Innovation*.

Chapter 15
Makerism

Cooperate globally, make locally

Makerism is an economic system that does not yet exist. The term is ours. In Capitalism, those who own the means of production are rich but few. In Makerism, those who own the means of production are also rich but everyone owns their own means of production, so everyone's rich. Wealth isn't traded (much) but is inherently distributed because it is made (as needed) and consumed by its creators.

Compared to the present American middle class lifestyle, Makerism aims for:

- Higher quality computational resources
- Transportation via Personal Rapid Transit and other innovative solutions
- Many trips displaced by higher fidelity communications, hyperlocal manufacturing, drone delivery
- User controlled and executed preventative health care
- Smaller, more insulated houses
- More energy and water efficient appliances
- Less, but more useful, stuff, often more customized to meet our individual needs
- More recycling: for instance, melting plastic goods to feed back into 3D printers

The technology of Makerism

Makerism depends on advanced, multi-purpose microfactories that can manufacture the material goods that individuals need to live a healthy, comfortable life.

The relevant processes are: printing, growing, cutting, casting, folding and assembling. Computers automate the machines that run these processes, taking the time and drudgery out of making. A good book on the many different kinds of machines and modern techniques for making is [Lipson 2012]. The forerunners of these microfactories are today's 3D printers, personal Subtractive manufacturing (called CNC) machines, laser cutters, aeroponic gardens and solar cells.

The typical, inexpensive 3d printer of 2018 is about a cubic foot in size, costing about $1K and can print plastic objects about a 6 inch cube in size. It has an extruder head that squeezes out molten plastic. By moving in the two horizontal dimensions, it can print a layer, then move up and print another layer on top. Printing tiny objects can take minutes whereas larger ones can take hours.

What things need to be made by the microfactories?

The most revolutionary thing a microfactory should be able to make is a copy of itself. This will include structural parts of plastic and metal, fasteners like nuts and bolts, electrical conductors for circuits and motors. Batteries and solar cells (or wind turbines) for power generation are also necessary. High resolution integrated circuits are a particular challenge, but progress on all of these fronts is continuing.

It is unlikely that a single-design microfactory will be the best for producing everything, especially those things that individuals need lots of. Number one in that category is food. We anticipate super-efficient microfarms using aeroponics (grown in air on a mesh) to supply fruits and vegetables à la [OpenAg 2017]. Aquaponic fish farming may also be used along with algae-growing microfarms, particularly good at supplying protein, fats, oils and vitamins. Those oils can be turned into bio-plastic and used as raw materials for our 3D printer. Textile producing machines are likely another specialty, for example [Kniterate 2016].

Larger, less commonly needed machines for furniture and house printing might be "neighborhood owned" [Costrel 2015]. Another strategy for large things is to print manageble-sized blocks and have a human assemble them. A chair might have 10 large parts and a bunch of fasteners. The essential parts of a house can be made with about 5K cubic-foot or less sized blocks. If it took a human an average of 5 minutes to place and secure each one, that's 417 hours, or a summer's worth of work for one person. Trading a summer for 30 years of mortgage payments sounds pretty attractive, especially if most other jobs have also been automated out of existence. With more advanced robots we can significantly shorten that summer of work.

The economics of Makerism

When a 3D printer can copy itself, its cost is merely time and raw materials. Time becomes not so expensive because as printers can copy themselves, you can have more than one of them and they can print in parallel. Raw materials can be expensive, but most things you want are made out of rather common elements: hydrogen, carbon, nitrogen and oxygen all of which exist in air. Dirt, leaves and other biological waste can form feedstock. Bioplastics can be made from algae which can be grown in a microfarm mentioned above. Aluminum is harder to come by but devices for extracting aluminum from dirt on the cheap have been designed along with many other tools for producing fundamental goods [Enigmatic 2016].

Direct recycling of unused items, especially plastic, is easy. Just melt it down and start over. With parameterized design, goods can be built with different material sets appropriate to the location of manufacture. The ability to make solar cells or wind turbines gets rid of electric bills.

Why transition from Capitalism to Makerism?

First, Makerism has the potential to be much more efficient than Capitalism. Roughy 3/4ths of a retail good's price is not manufacturing. Financing, labor and its management, transportation, distribution, warehousing, advertising, sales and taxes, all have middlemen with their fingers in the till. Under Makerism, all that disappears. You make what you need, period. With advanced printers, this will not be hard.

Second, being able to customize/innovate by "making your own" lets you tailor your possessions to fit, yielding a far greater variety than even today's massive malls and on-line stores.

Third, people just like to make stuff. They like to make it easy to make stuff, so they like to make tools. Unlike making software, making physical objects is more understandable, because it can take advantage of our evolved expertise at perceiving and manipulating physical objects.

Fourth, (this is the killer) we may well be pushed into Makerism by the collapse of Capitalism outlined in the chapter *Can Capitalism be saved?* [Ch. 14]. (Spoiler: Probably not!) Survival will encourage (to put it mildly) people becoming more self-reliant. Much greater self-reliance was the norm for most of human history, so we are not without precedent here.

But development of the technology will be harder from a position of poverty, so however much we can advance before jobs disappear, will greatly ease the transition. We expect this transition to take a decade or three, depending on how much effort is spent refining printers. The process has begun, though it has yet to have a major impact on the economy.

Most large scale change agents rely on "activism", i.e. organizing a large group of people to push for the change. In the late 1960's there was a large anti-war movement driven, not just by being against something, but by having the more positive goal of improving the world by promoting peace and love. These are worthy goals.

Unfortunately, empathy for others has proved to be insufficient. Merely making and distributing a huge quantity of 3D printers will not be sufficient either. But the combination of fine-grained distributed wealth creation as well as empathy may crack this age-old nut. Reducing scarcity will drive sharing and even greater wealth creation, in a positive feedback loop that promises to establish true civilization.

Makerism is not an end goal, it is a strategy. It is not an ideology. It is appropriate and efficient engineering. It does not need to get started. That's already happened. It does not need much more organization or motivation than its already got, nor does it need more "approval" from the status quo. However, the clock is ticking.

Meet the Makers

A Maker is a person that makes a variety of useful things. They do not sell the things they make, though they may give them away or perhaps trade with other makers. They enjoy giving away their designs for other makers to learn from and improve upon.

The antonym of "maker" is "consumer". A century ago and earlier, many more people were makers. A good example is a sheep herder who managed grazing fields (and thus manages solar energy), sheered sheep, spun yarn and knitted it into clothing. People "consumed" much of what they made.

The industrial revolution introduced "economies of scale", luring most factory workers into making a single kind of item, usually of a good they did not themselves use. Thus a worker made only a tiny percentage of what they actually consumed. But these processes also led to dis-economies of scale, where the added expenses are measured not in money, but in control, pollution, dissatisfaction, competition and alienation from other humans.

There is no official certification for becoming a maker, nor any official organization or trade union, nor will there ever be. There are a loosely organized series of Maker Faires hosted throughout the world. The first was in San Francisco in 2006. In 2015 there were 150 Maker Faires on every continent that people live. Millions of people have gone to Maker Faires. Some makers work at existing university and business research labs, while others work in basements or start-ups, many of which receive initial funding from crowdfunded websites like Kicksarter.

What does Makerism solve?

The key characteristics that define an economic system are: How wealth is created; and how wealth is distributed. Throughout most of human history, the creation of wealth has been the larger problem. Now our means of production are sufficient to create enough, but we have localized scarcity in much of the world. Makerism solves both the production of wealth and its distribution. For the first time in history, we have the potential to achieve a comfortable lifestyle for all.

According to [Mullainathan 2013], people that are poor, are literally better at managing money. They are experts at getting the most utility from a dollar. But this attention to detail comes at a cost, estimated to be 10 IQ points. It seems the concentration on scarcity limits ability to reason in other areas. Fortunately this is temporary: if the scarcity disappears, so does the IQ deficit. So the first problem we reduce by reducing scarcity is stupidity. Scarcity of nutrition, especially while young, causes a more permanent deficit in IQ. Poor overall healthcare affects both physical and mental well-being, so reducing healthcare scarcity offers multiple benefits.

A scarcity of security is a bit more complex. If you live in a low-income neighborhood, your stuff is more likely to be stolen. Besides all the above problems, "Physical insecurity often undermines opportunities for girls to benefit from quality education, good health and decent work, and prevents her from engaging meaningfully with, and benefiting from, society and the economy". More profoundly: "while poverty can be a cause of physical insecurity, physical insecurity also further perpetuates chronic poverty." [CPRC 2016]. Poverty causes poverty.

Humans have a deep sense of fairness. This causes all sorts of problems. If we're deciding how to cut the cake, should we give everyone the same portion, (with endless complaints about accuracy) or sick/hungrier/bigger people more, or what? The Makerism solution is, print more cakes so that everyone has as much as they want. We do not subscribe to the theory that people want an infinite amount of cake, especially if they are secure in the knowledge that they can always get more cake if they want it.

By solving scarcity, we solve property crimes. We solve most contention in divorce. Patents become moot. Most legal issues evaporate and with them the need for lawyers, whose primary tactic is to cause animosity to extract more wealth from their clients. With this reduced need for Justice and laws, one big government expense goes down.

Our government now expends a lot for shared infrastructure, but under Makerism, wealth will decrease the need for infrastructure further decreasing the requirements and cost of government and, its inherent corruption under Capitalism.

When is this going to happen?

These technologies are advancing more rapidly than any other technology today. The range of materials, resolution, speed of manufacturing, and ability to combine multiple technologies, are not yet good enough for Makerism. But innovations on these fronts are occurring surprisingly regularly.

The chapter *A day in the post-scarcity life* [Ch. 18] gives a bunch of specifics. The goal of the RepRap project is to make a printer that can print all of its own parts. The September 2015 version, named "Snappy" can print 73% of its parts [RepRap 2015]. We expect that the achievement of a practical self-printing machine will spark a takeoff in the ubiquity of personal manufacturing.

Civilization

We've dumped pretty hard on Capitalism. To be fair, historically, Capitalism has raised the standard of living for billions of people. But now, the very mechanisms by which Capitalism has created benefits, now act to limit those benefits.

A generally winning strategy to overcome these weaknesses of individuals is civilization. By amassing the knowledge and motivation of a large number of people, we have the potential to overcome the shortcomings of individuals and small groups. This has been a work-in-progress since humans began.

To be sure, humans have numerous limiting cognitive biases. Among the most relevant for our economy are: selfishness, immediate gratification, confusion in the face of complexity, risk management and fear, all of which impair our ability to reason. Most of humanity is out of the jungle, but the jungle is not out of humanity.

We outline a new kind of economic system called Makerism, that promises to elegantly solve wealth creation and distribution like no previous system could have. It's hard to improve other aspects of society, such as government, when scarcity pits us against one another. Makerism is therefore the enabling condition for civilization.

Chapter 16
Software Makerism

The concept of Maker doesn't only apply to physical objects. We believe an important part of Makerism is *Software Makerism*—the ability of ordinary people to make the software they need and want themselves, or in small groups.

We can't have hardware Makerism without some form of software Makerism. People have to learn the Computer-Aided Design tools that allow them to design things to print. Though perhaps most people will download designs from web sites like Thingiverse, there's an awful lot of designing to be done. (The web site Thingiverse had 880K items for free download in 2017. It is the largest, though not the only, site makers use for sharing designs.)

What we're talking about here is the ability for ordinary people to use the amazing procedural powers of modern computers, phones and other devices to help solve personal problems, and enhance personal creativity in people's lives.

This is called *end-user programming*. In order to maximize self-sufficiency and customization, the power of programming has to become accessible to everybody, not just a specialized professional class of people who have to learn esoteric languages. The easiest way we know of to learn the definitive reasoning required for rational governance is to learn how to program.

Currently, the production of software is largely done by large-scale industrial capitalism, companies like Google and Microsoft, mirroring the production of cars by companies like General Motors. These companies can only make software that supports processes performed identically by large numbers of people. Otherwise, there's no market for it, and they can't justify the expense.

Sure, there's plenty of software to help run small businesses, but nobody's going to write software specifically for Ahmed's Falafel Shop in Bay Ridge, Brooklyn. Ahmed wants to use the accounting terms he learned in Arabic back in Fooistan, which are a bit different. But then he's got to generate a report for his tax guy in English, so he can't just use the Arabic version of the accounting software. Sure, there are plenty of CAD systems, but they aren't much use for Walt's unique glass sculptures for his next art gallery show, where he melts photographs into flowing, curved vases.

The key to programming is language

But isn't programming a difficult, specialized skill that only science nerds with years of training can do? It doesn't have to be. The basic concepts of programming are already familiar to us from everyday life.

We all make decisions, so we're already familiar with "If ... then..." statements, what programmers call a *conditional*. We all have to-do lists, and other kinds of lists. Lists are one of the basic data structures used by programmers. When we fill out a form, we know to put our own name in the space that says, "Name", what programmers call a *variable*. We all seen recipes for cooking, which are composed of a sequence of steps. We all know how to do repetitive tasks, or some sort of *loop*. These are just the names computer scientists give to these concepts.

A programming language is a special language for talking about these kinds of concepts. Today's programming languages grew up out of mathematics and engineering languages, which is why they look so opaque and scary to the uninitiated. New programming languages and systems that are specifically designed to be more accessible to a wider audience will be a key technology for achieving Software Makerism.

It's a little bit like playing chess. The rules of chess aren't that hard to learn, and neither are the basic rules of programming. But of course, becoming highly skilled isn't easy. The best chess grandmasters can beat almost anybody, and you can spend your entire life learning chess strategy. But anybody can learn enough to play an enjoyable game of chess, and anybody should be able to learn to do simple programming.

The key strategic skill to learn to become a programmer is *procedural thinking*. That is, you have to learn to be able to describe a procedure you want the computer to follow, in enough detail that the computer will be able to know what to do in any situation it may find itself. People do this when they prepare a set of instructions for others to follow. A kitchen recipe is a program. A set of how-to instructions for a home repair task is a program.

There are programming languages designed especially for beginners, such as Logo [Papert 1993], Scratch [Resnick 2017], and ToonTalk [Kahn 2017]. These are much better for beginners than languages designed for professional programmers such as Java, C++, and Python. But even languages like Scratch are still much, much too hard for widespread use. The beginners' languages can't manage complex programs. The pro languages are too hard to learn and too error-prone for most people.

But maybe we don't need programming languages at all. A radical idea (and the topic of some of our research [Lieberman 2005]) is to make it possible to program just in (spoken or written) English. One of the original programming languages, Cobol, had the idea of making programs read as much as possible like English. It was the most successful programming language for business applications of its time. But it was only English-like on the surface; it had strict rules for a correct way to express something, just like other programming languages.

Now, advances in natural language understanding may make it possible to understand a wider range of variations of natural speech, as do conversational agents like Siri. The challenge is: can we understand *enough* to be able to do programming (which Siri can't)? It still won't get us out of the problem of *procedural thinking*—expressing programs in precise enough form that the computer can take care of all the contingencies a program will experience. But it could vastly increase the accessibility of programming to most people.

Some people are visual thinkers rather than verbal, such as artists and designers. These people usually find programming difficult, based, as it is, on words. Scratch and Toontalk are visual programming languages, so visual thinkers often find them more congenial. But they still use abstract representations of programs, and their visual vocabulary is extremely limited compared to what appears in art and design. So new programming languages that exploit a wide range of visualizations are needed. It's also possible to have programming languages based on other modalities, such as gesture, like the one depicted by Tom Cruise in the movie *Minority Report* [Spielberg 2002].

Sometimes, instead of programming by telling the computer what to do as a set of rules, it would be better to just show it by example. People learn best by example. If you do your taxes one year, the process you'd go through for next year's taxes isn't that different, except maybe for a few details. A solution to that is called Programming by Example [Lieberman 2001]. You show the computer a sequence of steps on a concrete example, and it remembers them. Then it creates a program that can be used on different examples (like next year's tax return) in the future, following the same pattern. It can generalize the program to be more flexible, so that it may apply to situations where the new examples are only similar to the old ones in some way.

Writing a program in languages like Scratch means you have to start from (excuse us) scratch. You can't use the vast capabilities of all the other programs you have already. Programming by Example is a solution to that, because you should be able to demonstrate a procedure using any program you already have.

Another important aspect for demystifying programming, ignored by most beginning and expert programming systems alike, is debugging. Beginners are shocked to find out that almost all programs they write don't work the first time. Most systems leave the programmer pretty much on their own to figure out what happened. Debugging tools in most programming environments haven't substantially changed in decades. Many aspiring programmers quickly give up when they see how hard debugging is. But there are some great ideas about how to make debugging easier, such as a debugger that lets you run your program forward and backward at different levels of detail [Lieberman 1997].

We'll need a considerable amount of innovation in programming languages and environments before we can truly get to Software Makerism. A good programming environment not only helps you manage complexity. It also teaches you about the language, your program, and most importantly, how to clarify your own thought process.

Hackathons and the Maker movement

You can see an inkling of the democratization of programming in the current popularity of "hackathons"—social events where people get together to write programs on particular themes. They brainstorm together, share ideas, make presentations, eat pizza, stay up late, and build community. The social atmosphere of these is wonderful.

It's just a shame that the programming tools used in most hackathons are still so difficult that full participation in a hackathon is generally limited to people who are already fluent in programming. And even then, the duration of such events, usually a day or weekend, makes it hard to accomplish substantial work. Many current hackathons are exploitive, run by companies to get free work from people, steal ideas for their products, or scout out potential employees. The prizes given out by hackathon organizers serve to encourage a competitive rather than cooperative atmosphere. But as programming gets better, events like hackathons will become a way that more and more people will drive the Software Makerism movement.

When all these new technologies come into play, we expect that writing your own programs and sharing them with friends will become commonplace. As in hardware Makerism, people will want to share designs and contribute to the community. Software won't have annoying ads, license agreements, privacy violations, etc. since it will be at the service of users rather than serve the commercial interests of the companies.

As Software Makerism rises, software companies will get increasingly disintermediated, just as we believe hardware companies will be disintermediated by 3D printing. For example, we are writing this book using free, open-source software rather than a commercial program; not because we don't have to pay for it, but because it appears to meet our needs better.

We were originally going to end this chapter with the line:

> *Programming is too important to be left to programmers.*

But upon reflection, we thought, why should we assume that programmers are a small, specialized group to be "left to"? Instead, we'll say:

> *Programming is so important that everybody needs to be a programmer.*

Chapter 17
Artificial Intelligence: Not the son of Frankenstein

A theme of this book is how advances in technology hold the potential for improving society. Perhaps no technology is more consequential than Artificial Intelligence—software that models human intelligence. But perhaps also, no technology is more controversial than artificial intelligence.

To date, society only improves through human action, and the time and energy and resources of people are necessarily limited. Artificial intelligence holds the possibility of amplifying human potential, putting an end to scarcity once and for all.

But people worry: will artificially intelligent computers become hostile to people and kill us all? Today, peoples' livelihood is tied directly to their (perceived) individual economic productivity, so will computers make all the jobs obsolete? Will computers do all the hard stuff and will we become bored and lazy? A vast majority of depictions of AI, whether in science fiction or from brow-furrowing social critics, rehash the Frankenstein theme.

In Frankenstein narrative, tailor-made for media sensationalism, scientists are toying with powerful forces beyond their comprehension, with little regard for humanity, and it is bound to go out of control and go berserk.

The authors of this book have devoted much of our lives to research in artificial intelligence. So we feel like we are at least qualified to suggest an answer to the question, "Will AI's eat us for breakfast?". The short answer: No. Don't panic.

Will intelligent machines get along with human beings? Our answer to that question is exactly the same as our answer to the question of whether, in the future, human beings will get along with each other. You already know what we're going to say:

Why can't we all just get along?

We have no choice, in some sense, but to build artificial intelligence in our own image. Principles for creating AI programs derive from generalizing our own experience in problem solving. The standard of whether AI programs succeed depends on comparing them to human behavior. The better they get, the more like us they will be.

So the answer to whether AI is a positive or negative development depends on whether or not you are optimistic about human nature. We do admit that there's some room for disagreement there, but we confess that we come down firmly on the optimistic side.

When people ask us if machines that are more intelligent than people pose a danger to humanity, we ask them back: "Do you think that *people* who are more intelligent than you pose a danger to humanity?". Let's not frame the question in terms of humans vs. machines, but this way: How do we have a society that is composed of various agents, some of whom are more intelligent than others?

Trouble is, our present political and economic systems are not so good at answering that question, either. All they can offer us is endless competition for power between humans. If the competition is about intelligence, then what you'll get is: The less intelligent will become slaves of the more intelligent. That's a lousy answer. In that case, yes, computers will conquer humans. But there's a better way.

Why can't we all, humans and machines, just get along? We want to have a society where, even if people have varying intelligence, everybody is valued. The more intelligent may have more to contribute in certain respects, but that doesn't mean that they shouldn't cooperate with the less intelligent.

If the agents (human or machine) are truly intelligent, they'll also have the *social intelligence* of understanding that cooperation is best for everyone, including themselves. As AI advances, we're increasingly learning that general intelligence really *requires* social and emotional intelligence. Work is advancing on how to model these different kinds of intelligence [Goleman 1995] [Mason 2008] [Minsky 2006] [Picard 2000]. We think emotionally intelligent AIs won't be hostile AIs.

The problem with most of the Frankenstein scenarios is that they assume that there will be technological progress sufficient to create an AI, but they don't posit much future progress in social intelligence. Our view is that there has been considerable progress in human relations in the last few hundred years.

For example, democracy (despite all its flaws) is a vast improvement over feudalism. It would have been hard for a citizen of a feudal society to envision modern democracy. As AI researchers, we realize that cracking the problem of intelligence is still far away. So we think it likely that future progress in social sciences will address problems of interpersonal conflict that now seem insoluble.

We hope that neuroscience and psychology will have advanced to the point where an insatiable lust for power will be viewed as the mental illness that it is, and we will learn how to create an AI where the risk of it developing such an obsession is vanishingly low. We do think the social critics are right in saying that there ought to be much more research in how to make AI more safe and controllable, just like we have research today in automobile safety or aviation safety.

In computer science, we call the problem of avoiding and mitigating negative outcomes, a debugging problem. Research in debugging, program verification, controllability, etc. is vital, yet very little research on this takes place, compared to research in optimizing computers to run faster or building new features.

We have been tireless in advocating research in debugging [Lieberman 1997], but our appeal has mostly fallen on deaf ears. Studies show that programmers spend, on average, more than half their time debugging, but the debugging tools, if any, provided by most interactive programming environments have changed little since the dawn of computers. No wonder computers continually seem like they're on the verge of going berserk. AI technology needs to be enlisted to help with the debugging problem.

In short, we feel like there are two possible futures for humanity: in one, the current status quo based on competition and scarcity, continues. In that case, AI will become a weapon of one or another of the competitive factions, and is bound to run amok sooner or later just like the prophets of doom fear.

It's too easy a game to come up with how doomsday might occur. In fact, we already have loose in our society, nonhuman machines that have more intelligence than any individual human being, are single-mindedly dedicated to amassing resources for themselves, and don't necessarily have the well-being of humans at heart. They're called corporations, governments, and religions. Some government trying to "defend itself" might invent a hyper-aggressive AI that goes out of control or is imitated and bested by its adversaries. Some company trying to "increase profits" might create an AI that is smart enough to trick people out of all of their money. Some apocalyptic religion might create an AI to hasten the apocalypse.

The real problem here is the adversarial relationship. In the future, we might not even *have* the social structures of militaristic governments, greedy corporations, organized religions, and criminal gangs that might be motivated to create malevolent AIs. We'd better not. Let's give ourselves the benefit of the doubt.

In the more optimistic scenario, people finally learn how to cooperate constructively with one another on a large scale. AI will help people arrive at this point, in part through AI systems for collaborative problem-solving, optimization, negotiation and conflict resolution. We don't believe that destructive competition and conflict are innate; rather, they are a response that is forced by scarcity and fear. AI's will solve the material scarcity problem, and education, psychology, and other human sciences (assisted by AI), the fear.

The reason that research in AI is so important is that it is the best route for bringing the positive scenario about. In that case, AI's won't want to destroy or "take over" humanity, like a James Bond movie villain. The AIs, too, will understand the wisdom of cooperation.

Chapter 18
A day in the post-scarcity life

The below fiction is set in the near future. It describes how a community infused with advanced maker tools might function on a daily basis. Most of the technology needed to support this scenario does not yet exist, but research has been started. The references are to those existing technologies, to help you assess how close we are to realizing this vision. Whether it will be a decade, a century, or never, is up to us.

Jack and Jill live in a house with their two teenage children, Johnny and Joanie. At Jack and Jill's wedding, the guests brought food to eat, some original poems and even a song composed for the occasion. They did not bring conventional wedding presents since Jack and Jill could make all their silverware, plates, [I.Materialize 2017] and wooden salad bowls [3dPrintingFromScratch 2017] on their 3D printer.

We begin our "day in the life" with Jill getting out of bed. "You know" she says to Jack, "my pillowcase is getting pretty worn out." Jack replies "I'll take it with me when I go down to the recycling center this morning to work on a new signaling system Joe has designed." Jack doesn't have a job, so he has time for such new fashioned "barn raisings", especially for projects as important as the recycling center.

Jack takes the old pillow case to the recycling center. You can recycle glass and aluminum there, but those quantities are *much* smaller than during scarcity, because beverages are made at home and don't need "carry away" containers. Plastic is a more common item to recycle. Failed prints and old toys make up much of it. Most people have small plastic recyclers in their homes, a la [Filabot 2017], but the center maintains one capable of recycling large pieces of plastic. It's 3 feet long. 2 feet wide and 1 foot deep.

The post scarcity society has a new perspective on recycling. First, everything that anyone ever thought of throwing out is potentially useful. This accounts for the inefficient clutter of scarcity societies. Next, knowing that you can print something when you need it helps you get rid of those things you "think you might need some day". Third, due to rapid innovation and free access to such designs, by the time you might need something that you had considered saving way-back-when, there's a good chance that an improved design will be available by the time you need that not-needed-now something.

Jack throws the pillow case in the hopper. The recycler takes apart the pillow case and makes "new" thread out of it, winding it onto large spools. Anyone can come to the recycling center and pick up these spools. As it turns out, houses have roughly a constant amount of textiles in them. Clothes, sheets, blankets, tablecloths, towels, curtains, upholstery and rugs all get recycled.

Joe tells Jack about his new invention. Joe maintains the "threads" recycling center but doesn't want to be there all the time. He does take pride in it having a 99% up time. But that means he or his assistant need to be notified as soon as there's a problem. The new button that Jack is helping him install automatically calls him, and if he doesn't answer, calls his assistant. If *he* doesn't answer, it allows the caller to record a message about a problem at the recycling center. The new switch required a bit of wiring [Bullis 2015] and an old cell phone. The cell phone, like all cell phones in the community, were made on printers [Google 2017 Ara].

They have no "service plan" or monthly bills. The phones communicate with each other, forwarding communications appropriately via a mesh network [Metz 2014]. The recycling center and people's homes are powered by solar cells that they print [Milsaps 2014]. Water is collected off of roofs (or even the air) and filtered with printed collectors and filters [Simon 2015b] or even a solar powered "still" [Simon 2014a]. Thus there are no water bills. Sewer "blackwater" is not created. Composting toilets a la [Seaparett 2017] don't use water and make it easy to safely handle human waste. Greywater gardens or recycling, or miserly appliances like this washing machine [Xeros 2017], handle the other waste-water. There are no sewer bills, or pipes to maintain.

After installing the new switch, Jack selects a spool of thread to make a new pillow case from. The recycling center supplies 6 different diameters of cotton and nylon thread, in near-white and near-black. Jack picks up a half pound cotton near-white spool and brings it home to his knitting machine [Salomone 2014]. He takes a picture of the matching pillowcase and sends the color to the dye dispenser, picks the design from "in-house items" and lets the knitter create. It's ironic that weaving started the industrial revolution but textiles have been one of the more difficult products to "print". More complex textiles need an automated sewing machine such as [Softwear 2017] and [Sewbo 2017].

Once home, Jack makes lunch for himself by picking a salad from his aeroponic garden [OpenAg 2017], and nets a fish from his aquaponic pond [Tilapia Vita 2017]. The kids prefer drinking lunch a la [Soylent 2017] whose powder comes partially from their algae microfarm. (Under the right conditions, algae can double its biomass in 24 hours. It supplies protein, carbohydrates, and oils. The oils can also be used to make plastic that can be printed into any shape. Plants can help make additional raw materials [Tampi 2015].

Johnny is filling out applications for college. There's no test scores to report, (his high school didn't give grades or standardized tests, and college doesn't require them.) only a "portfolio" of things he's made. Johnny is applying to the nearby "Maker U". A degree from Maker U isn't, from a capitalist perspective, economically significant. But there is some social status and just plain self-sufficient practicality to be gained. Plus for makers like Johnny, it'll be fun. Everyone's major is "making" with required courses including re-engineering designs for local raw materials [Makerbot 2017]. Designs now come in "flavors" depending on what's readily available in your region.

Johnny's minor will be in generating electricity [Molitch 2014]. He cut his teeth on an "erector set" for 3D printers and CNC machines [Uberblox 2017]. His portfolio contains a wind turbine he printed last month [Simon 2015d]. Even though it is nearby, Johnny intends to live on campus. (Kids still want to get away from their parents!) There's no tuition or fees and it's not tax-payer supported. It is self-sufficient. The Maker campuses build their own buildings using large 3D cement printers and large format "subtractive" technologies [BuildYourCNC 2017]. Another style uses smart lego-block like components that contain structure, insulation, wiring and plumbing [Simon 2016]. Food is grown in larger-than-home sized hydroponic units [Freightfarms 2017]

Health care is mostly provided on-site via easy to use "personal" sensors [Qualcomm 2016] and on-line physicians. The optical department determines eyeglass prescriptions using [Eyenetra 2017] then prints the glasses on the spot [Luxexcel 2017] Some body parts can even be printed, including bones [Milkert 2015].

The professors live there just like the students do. They get no salary, but, like the students, have comfortable living quarters, all the food they can eat, and plenty of eager students to teach how to build things. There are some lectures, but many more hands-on activities and interactive discussions than universities from the Industrial era.

Johnny's sister Joanie is already enrolled in college but lives at home. She takes classes on the Internet [Edx 2017]. Her concentration area is mental health, so her classes contain lots of video of patients. Twice a month, she works a weekend at a clinic in the big city. There's no salary, but it does go on her resume. You need a certificate and a resume to get even salary-less jobs in the prestigious mental health care field. Lawyers don't exist, but doctors do. Unlike the old days, doctors don't waste time filling out insurance forms. You walk into a clinic and get fixed. ID is required just to look up your medical record, not for payment.

Jill accidentally drops her running hair dryer into a sink full of water. Whoops, time for a new hair dryer. She goes to [Makerbot 2017]. Most of the content of the Internet is written by people that just like to write and give it away. Same turns out to be true for 3D designs. It is just some people's thing to design hair dryers, as it turns out. She reads a few reviews, and selects a model that is more energy-efficient than the one she busted. Jill downloads the design.

The original design is white, but Jill's favorite color is baby blue. She customizes it to her liking, something not easily done without your own printer. Heck, it needs a little electric fan motor that her printer can't make. Turns out there's a motor specialist in the next town over. Might he accept a scarf Jill recently designed? His wife's birthday is next week. "As long as it's got purple in it" is his response. Jill schedules a drone for pick up at her house in 1 hour, interrupts the pillow case job, then starts the printing of the scarf.

Model airplane building interest hasn't faded over the years, it just employs 3D printing now [Ulanoff 2015]. When the drone arrives, she plugs it into the batteries that have been printed to hold charge from the solar panels [Krassenstein 2014] , then loads in the scarf, and schedules it for take-off when it's got enough charge to fly the 5 miles to the motor-maker's house. In a half hour, the drone returns, drops off the motor in her "drop box" and flies off. As printers get more capable, these kinds of transactions become less necessary. Fry's Law: printer improvement decreases trade.

The hair dryer has plastic parts *and* the motor, so it needs both a plastic printer and a pick-and-place machine [Buzz 2015]. Jill places the motor right next to the printer, checks that the design from the net is properly loaded, and hits "start". This is a little more time-consuming and complex than buying a hair dryer in a store used to be. But she doesn't have to drive to the store, doesn't have to have a job to pay for it, or commute, or file complex tax returns. She also gets to pick the best of a thousand, instead of the best of ten.

The price of a retail product during the Industrial era averaged 4 times the costs to manufacture it. There are many fewer "products per factory" in the Maker era due to the distributed, ubiquitous printers, but the products don't have to be scheduled, transported, warehoused, retailed, advertised, marketed, taxed, remaindered, make a profit, or pay the graft and lobbying costs of the previous era.

That evening there's a neighborhood government meeting. Anybody's allowed, but just like in the past, young people tend to be interested in other things. Adults are more likely to attend, and do so more than "town meetings" since they've got more time sans–job, and the boring fiscal stuff is missing. Jack and Jill get on their bikes [Lai 2011] and head on over to the house where this month's meeting is. Though it is called a government meeting, there's actually a lot less to govern post-scarcity.

Agenda item 1 is dealing with the teenagers that egged a resident's garage last Halloween. The town practices "Restorative Justice" [Justice 2017]. Ron and Rollo were caught on Mrs. Magillicutty's security cam peppering her garage with rotten eggs. They stunk and damaged the paint. Along with the victim and offenders, their family members were also attending.

Step 1 was an admission of guilt by Ron and Rollo.

Step 2, an apology by them to Mrs. Magillicutty.

Step 3, "sentencing". It is not intended to punish the offenders so much as "restore" the loss to the victim. After 15 minutes of deliberation, there's consensus on the boys' sentence. They would be required to research sustainable paint, mix it up, print some brushes and rollers (or a sprayer if they deemed that appropriate) and paint the garage.

Unspoken in front of the boys, but understood by the adults, was the real reason for choosing that "sentence". Ron had already shown an interest in Chemistry. This was an opportunity to advance the boys' education in making something useful for others. Such a task instills pride and garners praise from the community, which hopefully will keep them out of further trouble, and give them the respect of a contributing member of society.

At next month's meeting the boys show a brief film of their work, explain the new paint they discovered on line (that has fewer toxins than previous paints), and show off the sprayer-brush they invented to complete the job faster. They got a standing ovation.

Next up, Joe explains the new switch he and Jack installed at the threads recycling center. For such a short presentation and minor project, the applause at the end was atypically long. But the people weren't clapping for the switch. They were clapping for Joe, who had maintained the recycling center for years. Nearly every person in the room was wearing threads that at some time had been recycled at that center. Textiles are necessary for modern life and Joe was a hero for making that life possible all over town. Many people thanked Jack at that meeting for the shirt they were wearing. For Joe, this was better than the paycheck he used to get at Amalgamated Agglomeration. He also loved answering detailed questions about thread sizes for this and dyes for that. People's need to be listened to hasn't decreased since scarcity was cured.

During the break, Mark passed out his latest crop, mini-mangos. By grafting a mango branch onto an apricot tree, he created a ping-pong ball sized, rapid growing, mango to die for. A neighbor with a brown thumb convinces Mark to come to his house and help him get started growing the new fruit. Over the course of the next year, Mark's mini-mangos sweep the nation and win the farmer's prize at the Maker Academy Awards. The annual ceremony is virtual. Even the MC uses 3D Skype to introduce nominees and tell jokes [Kelion 2013]. The equivalent of "best actor" is "best body part". For "best picture" its "best material". Several years ago, 3D printed cartilage won both [Lalwani 2015].

The prize for winning "best material" is making the statues for all the other prizes in your new material. Since the material is innovative, the process and printer attachments for it won't yet be perfected, so the winner has to print out and drone-mail (dmail) them across the country. People sign up years in advance to be a refueling stop on the way, just because they like playing a part in the big event. Pieces that are more standard can be "teleported" between a scanner-printer pair [Simon 2015c].

The final agenda item of the evening is a new design from the Global Village Construction Set [Enigmatic 2016]. This group has identified the 50 different Industrial Machines that it takes to build a small, sustainable civilization with modern comforts. Their goal is to design and facilitate the making and using of said machines.

The new design is for their Aluminum Extractor. A representative of the neighboring town gave an overview of what a facility for extracting aluminum from clay would look like. He's proposing a joint project between the towns. Several people signed up for the meeting next week.

Jack and Jill bike home. There's no street lights (one less infrastructure to maintain and collect taxes for) but the extra bright 3D printed quantum dot LED's [Sullivan 2014] connected to their bike generators are more than sufficient illumination. She dons a strong and light 3D printed helmet made of carbon fiber. A protective eye shield was printed in polycarbonate.

Once home, Jill tells Jack what she's been up to. Her long time passion is women's rights in developing countries. Despite advances in the West, many regions of the Middle East and Africa still live in the primitive, war-causing culture of Capitalism.

One of the barriers to advancement is the inability of women to gain more wealth by having working 3D printers. The challenge is not so much the making of the parts, which can be produced by other printers, but their *assembly*. The instructions often mirror IKEA instruction manuals of old, characterized by blurry gray photos, ambiguous, broken-English micro-font text, and missing steps.

Jill's contribution has been to update these instructions for the intelligent phones, that even poor women tend to have. Their 3D displays have ambient light sensors used to dynamically adjust color fidelity. The resolution is high enough that you can count the threads on the smallest of bolts. Triple clicking a part makes it "life size" so that comparison with the actual part is trivial. The latest technique is to have a double tap with a fingernail on the virtual part make the sound that would be made if you did the same thing on the physical part, allowing a user to use acoustics to sense material. The "wine glass" demo is a favorite.

Thingiverse designs have been extended to incorporate not just multiple flavors for regional raw materials, but different assembly guides as well. These represent different languages, cultures and styles. Authors, independent of the designers, can add their own manual and compete for ratings. They aren't compensated, just like open source developers. Nonetheless, enough people (like Jill) consider it worthwhile. Studies show that high-tech manuals like Jill creates, really do boost wealth in proto-Makerism economies. It turns out that the transition away from Capitalism has just as much to do with software as hardware.

After putting on the new pillow case. Jack drifts off to sleep wondering what percentage of planets make the transition from Capitalism to Makerism. In the Earth's case, luck was involved, but the global human characteristic of "Yankee Ingenuity" carried the day.

This chapter is not a prediction, it's a possibility. Even if this possibility is realized, it will differ in detail. The transition from Capitalism to Makerism, if it happens at all, may not be smooth. But just how rough depends on how humanity plays its cards in the next 10 to 20 years. We hope the knowledge in this book will ease the transition.

For additional scenarios in this space, please read [Tanenbaum 2015].

Part 4
Can government help us get along?

Chapter 19
The beginning of history

When society was hunter-gatherers, there were tribal wars from time to time. But when humanity was sparse, people could escape war simply by moving to another place.

When agriculture was invented, population density increased, and people had to stay put. There was the danger that you could work the fields all summer, then somebody would come steal the harvest. Battles between tribes became zero-sum contests over the same piece of land.

So an army for "defense" was invented. All over Europe, people lived in walled cities which were defendable, then went out to work the fields.

The military realized that they had power over people, and started to abuse it. They realized they could extort resources from their own people and use their military prowess to enforce it. They invented the *protection racket*—different military groups extorted money from their people on the grounds that they were "protecting" people from the other military groups (who were exactly the same). This is the same business model as the Mafia, or corporations and governments today. They had to demonize the others to stir up their people (thereby inventing *Nationalism* and *Racism* [Ch. 7]).

Note that you have to have more than one "country" for this to work. That's why leaders wouldn't support one world government. The politics of fear only work if you have something to fear—and it can't be in your own group.

The US Bill of Rights and the UN Universal Declaration of Human Rights are pretty good documents—they lay out the consensus of what people generally believe governments should do for individuals. But the UN has failed to implement these ideas. Remember, it's called the United *Nations*—and not the United Citizens of Earth or something like that. That's because it's a club for *leaders* seeking power over citizens, not the citizens themselves. See *No Leaders* [Ch. 22] and *Reasonocracy* [Ch. 25] for our solutions.

Leaders enlisted religion to help, convincing people that the military leaders had "divine right" to rule from God, and you got kings, queens, etc. who demanded unquestioning obedience and strict hierarchy. We don't understand why Europe is so proud of their "royalty" history. Kings and queens were basically tinpot dictators and thieves whom we'd deplore today.

This lasted until the Industrial Revolution, when Capitalism, Communism, and modern Democracy were invented. But we haven't completely outgrown Feudalism. Today's corporate and government structures echo feudalistic structures. We call them Presidents and CEO's, instead of Kings and Queens. They wear suits and high heels instead of crowns and robes. They carry briefcases instead of swords. But it's the same idea.

Going forward, we need to figure out how to reduce dependence on "leaders" and hierarchies. We need to distribute both power and responsibility more widely. We need to have open, rational discussion and debate in the public sphere about the best way to organize win-win arrangements for everybody.

The chapter *Some of Us* [Ch. 8] shows some alternatives for structuring collaborative decision-making without necessarily resorting to hierarchies and carte-blanche leaders. It's just a feudal-holdover belief that "you've gotta have a boss" as the only way to do things. The productivity of automation and the Internet's facilitation of global discussion give us, for the first time in history, a practical means of getting beyond Feudalism.

Francis Fukuyama famously wrote a book proclaiming "The End of History" [Fukuyama 1992]. He argued that, with the end of the Cold War, US-style Democracy and neoliberal Capitalism have defeated all existing alternatives (by which he meant mainly Communism, and what remains of third-world Tribalism and Feudalism). He reached this conclusion by not thinking very hard about what the alternatives might be.

We think that, instead, we stand on the cusp of the "Beginning of History". We believe that the citizens of the future will regard the solution to the problem of providing material resources for the world's people, as the enabling condition for civilization. They will look the present era of war, poverty, and environmental destruction, as a barbaric, prehistoric world. Welcome to the beginning of history.

Chapter 20
Toward low-power government

The 2015 movie *Selma* [DuVernay 2015] is overtly about racism, but we found the theme of power more intriguing. A battle between LBJ and MLK emerges. Key to understanding this conflict is a quote attributed to FDR:

"I agree with you, I want to do it, now make me do it."

This was never said in the movie but both LBJ and MLK behaved toward each other as if it had been. We interpret this quote as meaning that the decision maker, all other things being equal, wants to perform the way his advisor suggests. However, a president has pressure from all sides. If the pressure and his wishes conflict, that could lead to inaction. But if all criteria for making a decision point in the same direction, you don't even have to quibble about the weight of those criteria because the direction is clear.

Churchill said *"democracy is the worst form of government, except for all the others"* The movie depicts a time in America's democracy in chaos: marches, divided public, even assassinations. You might conclude from this that democracy had failed, yet we suspect the movie authors would claim that "democracy is inherently messy" and the process ultimately worked.

LBJ sponsored the Voting Rights Act, which MLK had lobbied for seemingly unsuccessfully. We are left to believe that LBJ finally sponsored the Act not because MLK asked him to or he wanted to, but because he was made to do it.

Power forced LBJ's hand. It was not the power of the gun, a Republican, or even money, but rather the power of the people. Observe the similarity between dictatorships and democracy: Decisions are guided by power.

American democracy is not unusual. The deciding criteria in most, perhaps all, large states is power, regardless of the form of government claimed to exist. What's so devastating about Churchill's quote is that it implies that no better form of government than democracy is possible. At least the vast majority of people believe this. Those that know Churchill's statement will utter it to prove their point. But Churchill, like most, was not an inventor. Inventors aren't limited by what currently exists.

Power, violence and fear

Historically, power-based governments arose out of people trying to solve problems by violence. Unfortunately, we haven't completely grown out of that.

A society is composed of individuals with conflicting motivations. With strong enough desires and narrow enough focus, people use any means at their disposal to achieve their goals. (You could point to this as a *cognitive bias*.)

In the most extreme cases, this leads to the use of force to manipulate the behavior of others. Violence and the fear of it are effective at manipulating others, but detrimental to society as a whole, and even to the perpetrators in the long term.

Large organizations—government, corporations, militaries, religions, and organized crime—institutionalize forcing their will upon people by creating persistent *power structures*. These structures anoint certain individuals or groups as having higher status, and condemn the lower ranks to obedience. That way, decisions by those in power don't have to be discussed or justified, one by one.

Power is a strong motivator, but it's only one-dimensional. Complex decisions are multi-dimensional. The best solutions are chosen by considering many factors, not simply "power" [Keltner 2016].

In order to get citizens to accept power structures, they have to be made to believe that without the power structure, the chaos of lawlessness or takeover by another such organization would be worse. That is, they must motivate the citizens by fear.

We've seen this pattern before, in the Prisoner's Dilemma. The power structure of the cops presents the offer to the prisoners, using each prisoner's fear of the other prisoner to get that prisoner to work against their own interests, and in favor of the cops.

One of the problems with power as an organizing principle is that it's a zero-sum game. It's impossible to appease everybody by giving them power in a conventional sense. What we can do is give people power over their own lives and make it easier and more fruitful to cooperate with others. Our economic proposal of *Makerism* empowers individuals, taking away the fear of scarcity, and preventing wasteful infighting. On our small planet, all fighting is infighting.

With Wright [Wright 2000],we've argued that evolution selects for positive-sum games. The good news is that, over the long term, evolution weakly, but persistently, favors cooperation. So our controlling institutions have generally moved towards more favorable organizations, less and less solely dependent upon power.

The bad news is that the process is slow. In the next chapter, we'll recap some history.

We made one giant leap when US Democracy, Version 1.0, was released on July 4, 1776. More than two centuries later a *lot* has changed. Innovation has changed almost everything. Why should government be immune to significant improvement? With advances in psychology, statistics, the math of cooperation, information collection, dissemination and aggregation, we think it shouldn't.

Still, there are aspects of the US Constitution that deserve preservation. Written processes help clarify and solidify fairness. Articulating the goal of "general welfare", having representatives, and a clear process for how they are selected, are all good ideas. But with changes in most aspects of civilization and our increased knowledge of ourselves, we believe the strategy for achieving the ultimate goals of the US Constitution can be drastically improved.

So, what's the expected release date for Version 2.0? As hackers say when they don't really know, "Real Soon Now". But let's get working on it.

Low-power government

It's now worth asking the question,

Can we have government without power?

Well, maybe not completely. But we'd like to introduce the concept of *low-power government*. Like LED lighting, low-power government would be more efficient, and better for the environment as a whole. Less heat, more light.

When we say "low-power", we don't necessarily mean the *minimal government* advocated by conservatives and libertarians. They want to minimize *any* government activity, regardless of whether it makes sense or not. The appropriate size of government is a question that depends on what problems it is trying to solve and what the solution requires. In any event, it takes some arguing to justify any particular decision about what government should or shouldn't do. By low power, we simply mean that the answers should be decided by having rational debate and the constituency reaching consensus. Not by the decree of pre-ordained power relationships.

We can reduce the heat, caused by friction between rival factions competing for power. We can increase the light, produced by the light bulbs of ideas. We need to develop a process of developing and evaluating ideas for improving society that doesn't depend upon the heat of power struggles.

That's what government should be, a *process* for solving society's problems. We can have the conversation about what makes a good process, rather than who should be the boss. We need to move from Aristocracy and Bureaucracy, to *Reasonocracy* [Ch. 25].

Chapter 21
Government at war with itself, and you

The framers of the 1789 United States Constitution were some pretty smart dudes. They were trying to design a completely new government, and had the freedom to propose all sorts of innovations. They tried to rethink the problem from scratch.

We wish today's thinkers about government would adopt a problem-solving attitude and be as open to innovation as the framers were. Now, many details of government are so ossified with tradition that we can't change them and we've even forgotten why they were there in the first place. Remind me again why we vote on Tuesday?[3]

All the examples the framers had to work with at the time were either European feudalism, with its kings and queens and peasants; and third-world tribalism and chaos. The founders knew that ancient Greece had some form of Democracy, but it didn't survive, so many considered it to have failed.

Some 18th century Churchill might have disdained the US Democracy project by saying, "Feudalism is the worst form of government, except for all the others".

---

3   From Wikipedia: "In 1845, the United States was largely an agrarian society. Farmers often needed a full day to travel by horse-drawn vehicles to the county seat to vote. Tuesday was established as election day because it did not interfere with the Biblical Sabbath or with market day, which was on Wednesday in many towns."

Design criteria for government

Like any design problem, the framers started by thinking about what the design criteria were, and what they wanted to accomplish. Then they tried to design realistic mechanisms that would meet the criteria they established, taking into account human imperfection. Some of the design criteria were:

- They wanted everyday people to have a say in what their government would do. They wanted the government, generally, to do what the people wanted it to do.

- They wanted individuals to have rights, things that the government could not do to them.

- They even incorporated many elements of the scientific process: They wanted a variety of opinions to be expressed, their pros and cons openly debated.

- They wanted to have a public decision-making process, including feedback.

- They realized that neither centralized government nor fully distributed government was best, so they tried to design a federal system that incorporated both.

- … and others.

It was (mostly) awesome. It gave us > 200 years of stability, peace and prosperity, relative to many other places in the world. It was widely imitated, and those who did also enjoyed its benefits. Like any design, it also had bugs. Starting with "all men are created equal"—first of all, it didn't include black slaves who were men, nor people who aren't men.

Architectures for cooperation, and for competition

In our terms, we can say what the founders intended to do was to create an *architecture for cooperation*. This is what government should be. Though this was 200 years before the Prisoner's Dilemma, it's clear that they appreciated the benefits of cooperation. They did all they could to foster it.

They acted out of hope that the "better angels of human nature" would lead us to a cooperative society. Many of the bugs came about because, although they had hope, they also had fear. What did they fear?

At the time, Europe was ruled by kings and queens that had absolute power. The founders feared that whatever positions of power they established, like an elected President, could be twisted into a new kind of feudal aristocracy. The new government could, as it did in Europe, conduct a war against its citizens.

Many of the founders were soldiers, such as General George Washington. How do you stop runaway power? The conventional answer: Why, by fighting it with more power, of course. So, while they set out to make an architecture of cooperation between the citizens and government, much of what they wound up with, was in reality, an architecture for competition between branches of government.

Remember what we said was the cause of war? *Fear of war is a primary cause of war.* The US government is constantly at war with itself.

Of course, we don't call it that. Sometimes we call it *checks and balances.* The three branches of government—legislative, judicial and executive, act as checks and balances against each other, so that no one can dominate. The House and the Senate check and balance each other. The Democrats and Republicans check and balance each other. This actually works. In 200 years of American history, no one faction in government has dominated for very long.

But it works only in the same sense as opposing feudal powers act as checks and balances to each other—at the cost of perpetual war. And, the potential for instability, should the balance go awry. No wonder current US politics is polarized and gridlocked.

Our system is designed to be *adversarial* and *competitive.* Legislators argue for and against bills, and vote for and against bills. Presidents can sign or veto bills. Lawyers argue for or against cases in the Supreme Court. One of the most common phrases of politicians is "I'm fighting for you". (Actually, they're only fighting to get themselves elected.)

As we've seen, emphasis on competition makes cooperation increasingly unlikely. When was the last time you heard a senator on C-SPAN say,

"You know what, I just thought of a third option that's better for all of us than either of the Democrat or Republican bills."

Hopefully, cooperation sometimes occurs, even if the system doesn't encourage it. But mostly, the visible mechanisms of government are designed for fighting, not cooperation. What cooperation does take place has to happen in smoke-filled back rooms. Maybe anti-smoking laws are responsible for the breakdown of cooperation in government.

Don't get us wrong. We think that some checks and balances, as in the design of the US government, are indeed necessary. There *is* always a danger of some piece of the government getting too powerful.

But adversarial structures should be a last resort, only resorted to after everyone really, truly, made best efforts to cooperate and scarcity looms. The military or warmongering politicians *say* "War is a last resort" but they endorse explicit mechanisms and incentives for fighting. Given short shrift are concrete plans, techniques, and funding for the supposed first resorts of cooperation, negotiation and peace building. So, in a competitive environment, last resorts have a habit of turning into first resorts.

Is US Democracy fixable?

Is the US Constitution of 1789 still a good idea? We observe that the context of a constitution matters a great deal. So what's changed since then? Pretty much everything (except, amendments aside, the Constitution text itself). The framers could not possibly have anticipated the Internet, TV, phones, airplanes, assault weapons, super PACs, corporations, labor unions, modern political parties, etc.

Also, scale. Average annual government expenditures in the late 1700's were roughly $16M. Today they're about $4 Trillion, or 250K times as much. If you believe that "Power tends to corrupt, and absolute power corrupts absolutely, then the Federal government contains a quarter of a million times more "corruption power" than it did when the Constitution was written [Madison 1787].

Today many political pundits complain of the paralysis of Congress due to partisan fighting (Democrats vs Republicans). As articulated in this chapter, the Framers set up competition between government segments to guard against any one segment gaining too much power over the others. If everyone was in the government, if there was more idealism (as appeared to be the case in 1789) and if there were fewer resources at stake, perhaps this strategy could make rational decisions. But none of those conditions now apply.

In particular, most people are not in the government but remain a resource from which the government can extract wealth from (under threat of jail). Those not attracted to power are pretty much excluded from it, considering the difficulty of getting elected. Democratic idealism has come down to buying votes with the citizens' own money via special-interest politics.

What about fixes that involve amending the Constitution? The last amendment was made in 1992. It was introduced in 1789 for a "gestation period" of only two centuries. Well, it must be a pretty good amendment if they had 200 years to get it right, right? The amendment delays approved congressional salary changes from taking place until the next representatives election (never more than 2 years away) [Wikipedia 2016].

What's so great about this amendment? It protects sitting representatives from being accused of voting themselves a pay raise (immediately). But since the vast majority of congressmen who run are re-elected, its not much of an arms-length decision. Furthermore considering the several trillion dollars in the Federal budget, its not a monumental decision. Imagine how long it would take for one that enacted significant change?

So, unfortunately, we don't see much hope of fixing the fundamental problems of power-based government by incremental changes such as constitutional amendments or passing laws one-by-one. It'll take a redesign.

We titled this chapter, "Government at war with itself, and you". The branches of the government, and our representatives, are constantly fighting with each other, but, as we explain in the *Prisoner's Dilemma chapter* [Ch. 1], that's a "pretend war" sideshow.

The real war is the government vs. the citizens. While the government is supposed to "be the people", it often acts in a self-interested way, to the detriment of the citizens. At $4T a year used for blowing up far away lands to bailing out Wall Street to subsidizing large corporations, you could argue the citizens are losing the competition.

Chapter 22
No Leaders

In 1887, Lord Acton made a now-famous pronouncement: "Power tends to corrupt, and absolute power corrupts absolutely." Less famous but perhaps more insightful is what he said next: "Great men are almost always bad men" [Acton 1887]. In case you're not a history buff, this thesis has been confirmed in psychology lab experiments [Bendahana 2015].

The notion of "leadership" never gets questioned in our society. Differences between US Democracy and Communism lie in how leaders are chosen, but they both agree on having leaders in the first place.

As we argued in *The beginning of history* [Ch. 19], leaders are just a holdover from Feudalism. And Feudalism, rooted in making decisions by power and violence, isn't going to help us get to a positive future.

All leadership processes in large organizations have the same pattern. There's a zero-sum process of "selection of leaders" (be the criteria military victory, elections, or commercial success). Strict, multi-level hierarchy ranks people (military rank, government rank, caste, corporate ladder, first class and business class plane seats). Succession is self-perpetuating (hereditary for Feudalism, political parties, corporate boards). Then we grant the "divine right" to the leaders to make arbitrary decisions that lower ranks must obey without question.

Sometimes there are processes for removing leaders when they get out of hand. This helps to clean up after the worst excesses occur. But it doesn't help prevent them in the first place.

Meritocracy

Meritocracy is a heuristic that improves the selection process for leaders. Instead of choosing leaders by hereditary, personal connections, or raw power, the idea is to try to select leaders based on their ability to make decisions. You "try them out" first in low-stakes situations, then if they succeed, you can promote them to higher levels where they have more power and responsibility.

Not a bad idea. It advanced us from raw Feudalism, to today's Capitalism and Democracy. But it has some bugs. Like Feudalism, it's still power-based. After the leader is selected, we still grant the leader the right to make relatively unquestioned decisions. (Unless they're so egregiously bad that they trigger the leader removal procedures.)

We should certainly respect expertise. If someone has a track record of being intelligent and making good decisions, it pays to listen to what they have to say. But not to have it go unquestioned. Even the smartest people and their best ideas are improved by critical feedback.

The problem is that leaders are only human. So, no matter how good the selection process is, you get a mixed bag. Some people get high positions by luck, gaming the system, or by aggression, and their decisions are probably worse than average. Others may be smarter, in the sense of raw intelligence. But putting them in a defined "leadership role" may tempt them to use their intelligence for their own good and not for the people they're governing.

The concentration of power in such a small group is itself a tremendous temptation for corruption. Sometimes, simply the pressure of having so much dependent on a single person simply exceeds the capacity of any human being to cope with it.

No time?

One argument for "leaders" is that they are necessary to make time-sensitive decisions. These can crop up on the battlefield or in business, but generally speaking, important decisions don't need to be made quickly.

Even decisions about whether to go to war generally don't need to be made on an emergency basis, since troop buildups take months. Post-

WWII, the US Congress has largely abdicated its role in declaring war to the President, so Congress can claim that the decision was made in the heat of an emergency. This provides convenient cover for Congress, especially in the case that the war turns out badly. War always turns out badly.

We can often predict what kinds of important decisions are likely to need to be made quickly, then carefully construct *rules of engagement*. These are principles that give guidance for various kinds of time-sensitive decisions, to streamline the decision process when it needs to happen quickly. People can specialize in certain kinds of situations, and responsibilities can be distributed.

For the exceptions, that's what phones and other immediate communications technologies are for. Put the few decision makers on speed dial with priority rings on their cell phones. A mistake in bombing a village is forever. A consultation between several people can take minutes no matter where they are in the world. We needn't continue to act like electricity has yet to be invented.

No President?

We contend that a major effect of having a head of state is increased chance of war. Commanders-in-chief are just plain prone to exercising their might because the process for selecting heads of state also happens to select for aggressive personalities. If heads of state were pacifists, how many wars would have been averted?

The politics of fear encourages presidents to cause war. It works like this: Fear is the strongest motivator. So if a president tells the public that "those guys are deadly", the public gets upset. Then the president provides a solution for the fear he caused by saying "Let's bomb the bad guys". Then he wraps himself in the flag and magically he gets support for being a true patriot. After 5 years of war and a bunch of citizens are killed, the public starts questioning the wisdom of that pointless war in the first place. But by then the election is over, or term limits kicks in, and the president is safe from public sentiment. If a president doesn't seize the opportunity to promote fear, a competitor will and likely win the next election. This is just one of the ways Democracy promotes war.

Even in ostensibly pacifist countries like Japan, pressure from a President can increase the possibility of war. Japanese President Shinzo Abe has proposed to "reinterpret" Japan's pacifist constitution, [Blum 2014] which outlaws war as a means of settling international disputes. You can guess what "reinterpret" might mean. Despite majority opposition from the Japanese people, and its potential to pave the way for the next Pearl Harbor, "top US officials" encouraged it.

A drastic solution would be to not have heads of state. But without other changes, having no President would put a lot of undeserved responsibility on the Congress. In US Democracy, legislators are pro war because they are paid to be by the Military Industrial Complex. Or they feel the need to use the politics of fear to get re-elected. Or they believe that killing stops killing. We don't propose axing the president without changing Congress. The chapter on *Reasonocracy* [Ch. 25] will supply more details.

No bias?

Another problem with having leaders is that they often have the power to appoint or personally influence the selection of other leaders. Due to the cognitive bias of *homophily* [McPherson 2001], where people tend to select other people like themselves, it can amplify racial, ethnic, sexual, and other kinds of discrimination. Male leaders can be biased towards selecting other males; white leaders select other white people.

Recently, after the impeachment of President Dilma Rousseff of Brazil, her white male VP, Michel Temer, appointed, for the first time in that country, a cabinet consisting solely of white males, causing an uproar in this highly multicultural country [Koren 2016].

[Pellissier 2011] explores the notion that having only women in government would decrease war. The article and the numerous comments following it make interesting reading about human nature, genetics and culture. Regardless, a more equitable gender balance is likely to improve decision making.

In the chapter on *Reasonocracy* [Ch. 25], we suggest that representatives be chosen randomly, like jury duty. Random selection pretty much guarantees we'll have a representative legislature. There will be roughly the same number of men as women. We'll get a proportional number of gays, including closet gays. We'll even get a proportional number of some minority that most do not even consider a minority. We also get rid of political parties, one of the worst features of US Democracy, because they prevent politicians from thinking for themselves.

No psychopaths?

A psychopath is someone who lacks empathy and is good at hiding that fact. Like most of us, they strive to get ahead. Unlike most of us, they are not self-constrained by treating others fairly and so have an advantage. Many think this results in a higher percentage of psychopaths becoming leaders than are in the population at large, which is maybe 2% [Freeman 2012].

Psychopaths tend to make poor decisions for those they oversee. One theory holds that villages of old, ostracized, or even killed, their psychopaths. Larger civilizations don't do that. One reason is that its hard to detect psychopaths without extended intimate contact because they are so good at superficially befriending people (in order to take advantage of them). In a larger society, its easier to "move on" or "fly under the radar" of most people, which isn't true in a small village. We may be inadvertently creating a higher percentage of psychopaths due to modern society's rules (promoting and voting vs killing the unwilling).

A leader can cause an outsized amount of damage to a society. Restricting psychopaths from certain jobs (police, legal, political, managers) would probably overall be a good thing. There are tests for psychopaths, though we should be extremely careful how we apply them. This is an area that could use a lot more research.

Ideas should lead, not people

The alternative to having "leaders" is to make our deliberation and decision-making processes be about ideas, not about people. Let's have ideas compete with each other, not people competing with each other. Even if the ideas should compete, the people should cooperate. The role of people is to consider a wide variety of ideas, and try to consider the pros and cons of each idea.

There's a big difference between competition between ideas and competition between people. Competition between ideas has an important role to play. Competition between mutually exclusive ideas helps you explore the ideas and their consequences, in parallel. In science, the purpose of doing experiments is to evaluate competing hypotheses, gather evidence, understand underlying principles, and of course, design new experiments.

However, when you tie individuals strongly to particular competing ideas, objectivity, balance and perspective tend to go out the window. If particular ideas win or lose a debate, that's great; either way everybody learns something. If we make people live or die by the success of their ideas in a debate, people get or lose jobs. There's tremendous incentive (read: Temptation) to push your idea by hook or by crook. Mostly by crook.

Even worse, we tie groups of people with similar views to each other (in so-called "political parties") so that deviation from one idea risks disloyalty to the group. Maybe you present a one-sided view; maybe you attack the opponent; maybe you dismiss contrary evidence; maybe you don't lie, but you exaggerate and distort. That doesn't enrich the debate. It makes it a game of Liar's Poker.

You might think that, if we say ideas should lead, then maybe we should have Direct Democracy, that is, voting directly for propositions, not people. While this eliminates the problems of leaders, it doesn't eliminate the problems of elections. We'll have more to say when we explore solutions in the *Reasonocracy* chapter [Ch. 25].

So we're with Bob Dylan: *Don't follow leaders. Watch your parking meters.*

Chapter 23
The trouble with voting

"The difference between theory and practice is:
In theory, there is no difference. In practice, there is."
— Yogi Berra

Ideal Democracy and US Democracy

Democracy is a kind of government. One of the problems in thinking about democracy is that there are two very different things you might mean when you use the word. In grade school, you learn about the theory of what we'll call *Ideal Democracy*; it is about people having self-determination, fairness, the "rule of law", etc.

In practice in the USA, a very different form of democracy takes place. We call this *US Democracy* to be clear. US Democracy is more about special interests and the power of a few, preserving the status quo. One might hope that democracy is a work-in-progress, moving from present US Democracy to Ideal Democracy. If such a move is taking place, its slow, and anything but smooth.

There are powerful *structural* reasons why this movement is difficult. By structural, we mean independent of individual actors. Part of the attraction of democracy is that it purports to be "fixable" by "voting the bastards out". Though theoretically possible, this has proved to be impractical. There are many levers that the incumbents can pull to insure their re-election. When those fail, new inductees become corrupted by their new-found power and join the status quo. Most remaining officials usually don't have the knowledge on how to select good ideas. Should there be anyone left, well, they're in such a minority that their voices will be suppressed or at least out-voted in committee.

To form a good government, you must have good decision makers and a good process by which those people make decisions. US Democracy has neither.

A vote against voting

There are few stronger beliefs in America than "voting is the best way to make decisions". We see this in national though local governments. We see this in clubs or even friends deciding where to go for dinner.

Before we disparage it, let's acknowledge a few good things about voting. First, we show some respect to the founding fathers. For the 1700's, voting was the state of the art. Given an expensive printing technology, no electricity, no telecommunications or computers, tallying up votes from white men (women and slaves weren't allowed to vote) pushed the limits of the "database technology" of the time. We can hardly fault the founders for not anticipating lobbying, TV ads, and SuperPACS. But 225 years later, we shouldn't be bogged down by their limitations.

The concept of each citizen having an equal say in decisions, i.e. "one man, one vote" has an appealing "fair sounding" ring to it. But even in the modern US, we don't conform to this fairness (i.e. the electoral college for President, 2 Senators from each state regardless of their population, gerrymandering in the House of Representatives ... hmm that covers 100% of the national elections). We've always had a two-party duopoly, with third parties systematically excluded. Despite the facts, we still like to pretend that voting is fair.

Voting also has a simplicity to tallying that is appealing to transparency and thus discourages corruption. In practice, this advantage too is, um, not so great. Witness the US 2000 presidential election, and the frequency that election results are challenged all over the world. Let's pretend, though, that we could implement voting as it is intended. Is it a good way to make decisions?

Voting is devoid of reasoning. Yes, each individual may reason on how to vote, but the process itself doesn't capture those reasons, it merely adds up the votes of the individuals. There are many different ways to calculate winners, with majority-rule being the most common.

We've even invented mathematically better voting schemes, which are nevertheless ignored at the national level. In civilized places like Cambridge, Massachusetts, elections are conducted with preferential voting. (Maine adopted this for most elections in 2016.) No political parties. You mark the candidates 1, 2, 3... in order of preference. No "wasted votes" [Fairvote 2016]. But all voting schemes ignore rationale.

One could argue that candidates inform voters about why they're the best. These pre-election reasons tend to have little to do with post-election behavior. But imagine, if you can, a politician keeping campaign promises. First there might be good reasons for the candidate violating their promises, so appealing as it may sound, this might not actually be what a voter would want. A politician is stuck between being accused of not keeping a campaign promise and not being flexible under new circumstances.

Second, democracies like the "one citizen, one vote" idea. So a highly informed citizen's vote is worth the same as someone who didn't even know who was running until they entered the voting place. However, restricting certain citizens from voting based on knowledge has been used for racial or other discrimination in the past.

Fake news

Having large numbers of poorly informed voters leaves them vulnerable to well funded dis-information campaigns. The two Koch brothers intended to spend nearly $1B on the US 2016 presidential election [Gold 2015] as did each of the Democratic and Republican political parties. (Due to dissatisfaction with the republican nominee, much of this money was diverted to state elections.) Dis-information in the form of "climate change denial" has been particularly prevalent [Romm 2015a].

The internet phenomenon of "Fake News" especially on social media, may well have been a significant influencer in the US 2016 presidential election. With advanced video editing technology, you can now manipulate facial expressions including what you look like when you're saying something [O'Keefe 2016a]. With advanced audio editing, its now possible to make a person say anything *in their own voice* [O'Keefe 2016b]. Thus video is now no longer solid evidence of reality any more than still pictures are post Photoshop.

Confirmation bias

It gets worse. People don't want the truth, they want what conforms to what they already believe. [Majoo 2016] says, "when confronted with diverse information choices, people rarely act like rational, civic-minded automatons. Instead, we are roiled by preconceptions and biases, and we usually do what feels easiest—we gorge on information that confirms our ideas, and we shun what does not."

"This creates an ecosystem in which the truth value of the information doesn't matter," said Walter Quattrociocchi, one of the study's authors. "All that matters is whether the information fits in your narrative."

And what about debunking fake news? "In many ways the debunking just reinforced the sense of alienation or outrage that people feel about the topic, and ultimately you've done more harm than good"

Democracy can't make good decisions without informed voters. Oddly, with all of our communications technology, its not simply that we have uninformed voters, it's that, at the present time, we lack the very *ability* to have informed voters.

How representative are representatives?

Not very, because its hard to win elections. Money helps, so not surprisingly, congress members in 2011 had an average wealth exceeding $7M [Ballotopedia 2017] and a median of about $1M. In 2013 the median US household wealth was about $69K [Census 2013] or about 1/15th that of the median congress member. Looking along other dimensions we have only 20% of congress female. The presidency with 1 black and 0 females (as of August 2016) doesn't reflect the citizenry.

What effect does this unrepresentative congress have on the decision making process? From [Gillens 14]: "Economic elites and organized groups representing business interests have substantial independent impacts on U.S. government policy, while average citizens and mass-based interest groups have little or no independent influence."

Well then, whose preferences do have an impact? "In the last 5 years alone, the 200 most politically active companies in the US spent $5.8 billion influencing our government with lobbying and campaign contributions. Those same companies got $4.4 trillion in taxpayer support—earning a return of 750 times their investment." "91% of the time the candidate with the most money wins the election." "2/3 of political donations come from just 0.2% of Americans." "Our elected officials spend 30-70% of their time in office fundraising for the next election." Congress literally doesn't care what you think. If a candidate wasted time learning what their citizens wanted, or thinking about making important government decisions, they'd be replaced by someone who didn't.

Corruption

Corruption has many other tentacles. Quantifying them all is practically impossible. However, Transparency International has made a decent stab at it by taking world surveys (of > 100K people in > 100 countries) on the perception of corruption. [Shah 2011] reports:

- 54% Think government is corrupt (Worldwide)

- 60% Think government is corrupt (UK)

- 62% Think government is corrupt (Mexico)

- 64% Think government is corrupt (USA)

- 25% Paid a bribe in the last year (Worldwide)

- 54% Think corruption has increased recently (Worldwide)

- 64% Think personal contacts needed to operate in the public sector (Worldwide)

And, surprisingly:

> "The democratic pillars of societies are viewed as the most corrupt. Around the world, political parties, the driving force of democracies, are perceived to be the most corrupt institution."

Globalization

Globalization appears to foster larger-scale international corruption. From [McCarthy 2014]:

> *"While increased international attention has helped move the anti-corruption agenda forward, globalization is responsible for an increasingly sophisticated form of corruption. We have to ask whether corruption-fighting solutions have kept pace with the integration of financial systems, global supply chains and multi-jurisdictional entities."*

From [Shah 2011]:

> *"Legalizing drugs, a system of taxation and regulation, comparable to that applied to tobacco and alcohol might do more to reduce corruption in the world than any other measure rich countries could take"*

> *"The business of obtaining oil and mineral concessions has always been conducive to the use of bribes, omissions, gifts, and favors, and remains so."*

> *"For multinationals, bribery enables companies to gain contracts ... These bribes are conservatively estimated to run to US$80 billion a year—roughly the amount that the UN believes is needed to eradicate global poverty."*

If not voting, then what?

Finally, although we've made harsh criticisms of voting, we'd like to stress that we're not entirely against voting. If you've got a group of people who *have* to make a decision, and have *irreconcilable* differences, then voting might be the least bad way to get out of the jam. So, we could say, "voting is a last resort" for making decisions. As we noted in the *War with Itself* chapter [Ch. 21], though, last resorts have a habit of turning into first resorts unless you really commit to the first resorts.

In the next few chapters, we'll make proposals for criteria and mechanisms of better government, though we don't have all the answers. For the moment, resist the temptation to criticize us immediately with, "Yeah, but how do we get from here to there?". Just see if you agree with us on what needs to be done. As long as we're proposing changes the status quo will reject, we might as well propose solutions with the scope to actually *solve* the problem.

Chapter 24
The process of Science

*"The binding force of science is its common language
extending rational thinking across borders, cultures and religions
to the benefit of all."*
—Ahmed Zewail, first Arab to win a science Nobel Prize.

Up to this point we've made all sorts of proclamations about how the fundamental institutions of our power-based governments are lousy. But anarchy doesn't work so well either. If we're going to get rid of such previously sacred ideas as leaders and voting, what are we going to replace them with?

Even before we get into specifics, its worth asking the question "Has any non-power based government ever worked?" We answer "probably not for long for a large population". We might not have examples of non-power-based *governments*, but we do have many examples in other parts of our society, of *governance processes* taking place in the absence of power relationships. Some of them were explored in the *Some of Us* chapter [Ch. 8]. What might we learn from them that can teach us lessons in how to redesign government?

Government is a *process*. We will not find our solutions in "electing the right people" or "better architecture in the capital". We need examples of winning *processes*. Exhibit A:

Science operates via consensus

One of the best examples of a community that operates primarily on cooperative consensus is one that your authors know best—the *scientific community*. Roughly speaking, Science, in one form or another, lies at the core of most of civilization's advances over the last 500 years or so. The revolutions in transportation, communications, information, medicine and numerous others all sprouted from Science. What's the magic?

Think about it. The scientific community involves hundreds of thousands of people throughout the world, tackling some of humanity's biggest problems. They produce life-changing breakthroughs on a regular basis. And with relatively modest expenditures, we might add.

Surprisingly, most of the organizational structures that business and government claim are necessary to get things done, are absent from science. There's no President of Science. No Supreme Court, no political parties, no CEOs, no boards of directors, no stockholders, no Pope. No hierarchical organization chart. Very little voting. If you've ever hung out with scientists, you know that debates can get heated. So how do these guys handle disagreements?

Cynics might counter that science indeed has competition and hierarchy, just like business and politics. There is the tenure track for professors, every bit as ranked as the military. There's competition for research grants, sometimes as cutthroat as marketplace competition.

But those activities are not what we're talking about. Those are not the activities that constitute science itself—rather, they are the business and politics of jobs in science, which is why they resemble business and politics in the rest of society. There certainly are occasional instances of scientific fraud and malpractice, trumpeted in the newspapers, but as a whole, by any measure, there's far less avarice and corruption in science than in business and politics.

Science 101

Science is a process for discovering, verifying and disseminating knowledge. Fundamentally, it is the process of using reasoning to discover the truth. It relies on careful documentation, reproducible experiments and reasoning, to derive explanations for a wide variety of phenomena. Let's dig a little deeper.

Just as there are no cans in a supermarket labeled simply "food", we've never heard of a college course labeled simply "Science". Science is not simply the sum of individual fields like Biology, Chemistry and Physics.

Does science Work?

Not for all subjects, not all the time, but pretty much yes. Over the long term, the process of Science tends to converge on answers that are useful in modeling phenomena, making more accurate predictions than non-scientific process and, in general helping us lead happier, more productive lives.

Admittedly, some products produced by Science-guided technology are, um, misguided. "Smart bomb" is an oxymoron. Refrigerators help, but the hole in the ozone they caused doesn't. There are a lot of "two steps forward, square root of two steps back", yet we can now cure diseases, fly thousands of miles in hours and acquire knowledge from all of humanity in seconds, largely do to Science and its spinoffs.

Generally problems that Science creates are problems Science can solve, given time, funding and freedom for creative minds to explore. Science is not all recluses in white lab coats.

The social process of Science

The real activity of science, exchanging ideas, takes place when scientists interact socially, in working groups, dinners, and conferences, when the bureaucrats and administrators are out of the room. The beauty of science is that scientists have developed a method of discourse that allows them to cooperate with each other to solve problems, rather than just to compete for money and status based on power.

Anybody can express an idea, from the lowliest student to the most distinguished professor, regardless of status, by submitting a paper (usually anonymously). The idea is evaluated by paper reviewers, who have expertise and interest in the subject matter, but no stake in the outcome of whether or not the paper is accepted. Conference organizers seek a diversity of viewpoints.

Individual scientists get to make up their own minds about the idea, without coercion. In the best case, other scientists will like your idea, and they will adopt it and build upon it in their own research. They credit the original idea with references in the paper. That's how ideas get adopted. Not by being blessed by an authority, or approved by a vote.

Scientists resort to competitive processes typically only when there's some sort of scarcity. You will recall from our discussion of evolutionary theory, that scarcity is a legitimate motive for competition. There are only so many speaking slots at a conference, or pages in a journal, so that's when voting or selection by authority takes place. And even so, there are typically feedback mechanisms in those cases, to assure that authority is not abused, or a vote too influential.

In science, everybody is assumed to be cooperating to discover scientific truth, rather than working for themselves in order to advocate a position, or chase incentives for personal gain. While complete objectivity is impossible, scientists are expected to make every effort to be objective in their presentation of ideas. It is established as a strong social norm, taught to students as part of their education. By and large, people play along. As a result, discourse in science is markedly different than the discourse that takes place in business and politics.

There are several roles that scientists play. There's the experimental scientist collecting data and running experiments. They design an experiment that can objectively determine if a hypothesis is correct, then analyze the results. Unlike most other fields, proving your hypothesis *wrong* can be every bit as valuable as validating your hypothesis.

Experimental work is published in narrow focus peer-review journals and conference proceedings. Next, some authors recognize the importance of a collection of work and make it available to a wider audience of scientists via survey papers. Other authors bring the more broadly agreed upon concepts to students via text books.

The more exciting ideas get published in Scientific American or Popular Science. The occasional breakthrough makes it all the way to daily newspapers. This layered dissemination of knowledge helps attract interdisciplinary ideas that can ultimately enhance the original work with ideas from a larger population.

When you present an idea in science, you have to

- Make a good-faith attempt to present both the reasons for and against the idea. You have to present opposing viewpoints fairly. Your paper will very likely be sent to reviewers with those viewpoints, who have to agree on the fairness of your characterization.

- You have to situate your work in the context of other relevant scientific work, and give credit to other people who have contributed. No scientific paper would be accepted if it lacks a references section.

- Any claims that you make, to attract people to pay attention to your work, must be backed up by the results of the paper. Work that fails to meet these criteria is simply ignored, rather than explicitly censored or punished.

So, unlike business and politics, there is practically no "marketing". No soundbite slogans, no focus groups, no polling, no bait-and-switch ads, no fine print.

Science tends to stay away from the practical near-term kinds of decisions that governments should make: Who gets how much money? What's illegal? *Where to Invade Next?* (also the title of a Michael Moore documentary). Which begs the question:

Can government utilize the process of science?

Broadly speaking our answer is: *yes*. But not unadulterated.

We can't take decades to resolve short-term issues. We can't be wishy-washy with "probably" kinds of decisions. We need fairness, compassion and nuance in solutions, not a table full of numbers. We need definitiveness where certainty of outcome is not guaranteed. Laws need to anticipate the future, not merely cover past known cases.

Each branch of science needs its own language to facilitate communication and thought of its members. So, too, does government by reason. Such a language is the core of our tool Justify, in the *Tools for Reasonocracy* chapter [Ch. 27].

Processes need skilled people to perform them. Scientists are knowledgable in their discipline and should be able to communicate that knowledge to a wider audience. Similarly, legislators in a government need to be skilled in reasoning and to be talented at expressing rationale to their fellow legislators and the public.

Science is a lot of work, but our investments in it have paid off more profitably than anything else in society. Given the ROI (return on investment) of Science, it's shameful that society is so underinvested. There aren't enough places for science students at top universities, and prominent scientists spend large percentages of their time begging for research funding from ignorant bureaucrats (we, unfortunately, speak from experience). Meanwhile, vast spending on the military, commercial competition, and other activities that actually have a negative impact upon society, go unquestioned.

Politics meets Science

As scientists, we are tired of hearing political people admonish us, "Scientists should get more involved in the political process".

Sure, politics would be improved if there was more input from scientists (and if political people really took science seriously). But scientists are reluctant to get involved in politics precisely because they can't stand the contentiousness, viciousness and irrationality that characterize public debate in the political world. If you want us to get involved, clean house first.

We'll turn that on its head and say, "Political people should get more involved in the scientific process". We're not saying that political people must learn the subject matter of science (although knowledge of a scientific field would certainly aid reasoning). But they should really pay more attention to how discourse about issues takes place in science, and use science as a model for how cooperation between large groups of people can be achieved.

The principles by which the scientific community organizes itself are far more reasonable than the political process of today's US Democracy. We'll build upon these principles to make a more specific proposal for government in the next chapter, on *Reasonocracy* [Ch. 25].

Chapter 25
Introduction to Reasonocracy

Government structures are notoriously difficult to architect. They are fraught with unintended consequences, unpredictable futures, and irrational implementations. Powerful individuals may seek to subvert the best of intentions at every step. But we've got to at least try; present US Democracy is so flawed that we're drifting at sea. We can't demand our rescue ship be water-tight before we launch.

Don't we already use reason in government?

Let's shift the emphasis away from tallying up votes made by individuals. In fact, let's shift the focus off of people altogether, and focus on the process by which decisions are made.

You might counter that we already use reason. You might say that voters reason who they should vote for. We therefore elect the most reasonable candidates, and those elected candidates reason with each other to come up with the best laws. This is the rationale for democracy. In theory, it sounds reasonable. Remember Yogi Berra's quote about the difference between theory and practice?

Mitch McConnell, head of the US Senate, said in 2010, "my number one priority is making sure President Obama's a one-term president" [Kessler 2017]. McConnell is not concerned with issues; he's concerned with a particular elected official. Its personal.

We've already pointed to the social processes of Science, which is based on logical reasoning, as an inspiration for how to organize large-scale cooperation. So we'll call our alternative, *Reasonocracy.*

We tend to think of government as a machine for manufacturing and enforcing laws. The design of our government is all about procedures for how deciders get chosen and how laws get voted on. It says little about why or how laws are introduced, how to evaluate whether a law is a good one or not, whether one proposal for a law is better than another, how to tell whether existing laws are working or not, whether there are solutions other than laws, etc.

Science has a lot of laws too, like the Law of Gravity. But scientists don't think their job is just to make laws. Their job is to solve problems. The Law of Gravity was introduced as a *solution* to the problem of explaining why things fall down.

Now, we've got the situation that the institutions of our society are falling down. So we need to rethink government as a vehicle for solving problems that require the cooperation of people on a national level.

Reason is inherently complicated. A few people sitting around a table for 15 or 20 minutes can generate hundreds of ideas. Relating those ideas coherently to each other is a task too complex for most people, unaided. Simply put, people need help in managing complexity.

US Democracy selects for decision makers that crave power. Those that don't are at a disadvantage as they waste resources on things that don't help get them elected. We are poor at picking leaders. Did you ever even hear of a class on this topic?

Direct Democracy

Consider the opposite extreme: suppose we do away with leaders and legislators and have "direct democracy"? In the Internet era, it would be technically trivial to set up a system whereby everybody got to vote on every issue. Would that result in better decisions getting made?

Actually, given the corruption of today's system, it might not be worse. But there would still be problems. Without a sufficiently educated electorate who were willing to spend the time to thoroughly investigate issues, one-shot votes wouldn't be good at maintaining a long-term perspective. California's propositions have shown that voters for propositions are just as easily swayed by big money to vote against their own self-interests as when they are voting for candidates.

One of the principal ways the scientific community solves problems is through dialog with each other. You can't have a meaningful conversation with 300 million people. So perhaps we shouldn't give up on the idea of having some sort of representative body.

Or maybe we *could* figure out how to make a 300-million person dialog practical, using technology. We'd certainly like to make it possible for anyone to submit an idea that could improve government. As in science, good ideas can come from anyone, regardless of rank. The problem with the write-your-congressperson way of doing it, is that the corrupt congressperson serves as a gatekeeper, ensuring that most of the good ideas will wind up in the circular file.

Instead, suppose we had some kind of national online forum for discussing issues. The sheer volume of ideas generated by this process could be its biggest problem. A group of professional "filterers" could help eliminate nonsense, redundancy and holes. We might allocate (perhaps several levels of) professional "editors" to help citizens formulate their ideas clearly. The expense of editors and filterers is justified by the value of lost great ideas in our current system.

The key is that the editors and filterers don't have a dog in the fight—their job is to facilitate discussion and consensus solutions, not to advocate for a particular position. This comprises a new channel for a nation to get the best minds of its citizens to contribute in an organized, productive way.

Editors could produce shorter, more focused discussions for the next level up. To ensure that it's not just the loudmouths who get a voice, some moderators could be assigned to interview people at random to get a sense of the "silent majority". The moderators would be professionals in facilitation, conflict resolution and consensus. Interested citizens could join at whatever level fits their interests and the effort they were willing to put in. And it would be a much better way for representatives to understand "what the people want" (and, crucially, *why*) than simply an electoral mandate.

But, ultimately, we still might want to have representatives, for the top level discussions, to determine when consensus has been achieved, and to take the occasional vote when it is needed. So, how do we choose them?

Random selection of representatives

Members of Congress need cash for their re-election campaign. They get it from special interests (typically wealthy businesses). Term limits is one solution, but that hasn't worked out well for presidents (see below), nor for California legislators [PPIC 2017]. Even if a legislator has reached his term limit, there are other offices to run for and thus other campaigns that need funding. Also, first time legislators will still feel beholden to their backers.

Suppose instead, legislators were chosen at random, like (voluntary) jury duty? Random selection pretty much guarantees having representative representatives. It still gets all voices heard with roughly a balance of power proportional to their weight in the population governed. But wouldn't we wind up with complete idiots as representatives? Well, we ask you, do you think voting screens out "complete idiots" [Rennix 2017]?

We might have some requirements: US citizen 21 or over, high school diploma, can pass a few tests of reading, writing and math. Actually, we're less concerned with idiots than the "selfish genius" types, which use their intelligence for self-interested motives. Since these types are a small percentage of the population, only a small percent would be chosen with a random selection method.

Second, we wouldn't put people into these situations cold turkey. Our next chapter, *Some days in the life of a Reasonocrat* [Ch. 26] dramatizes what this would be like. Admittedly, many people don't have skills in economics, reasoning, and cooperating in meetings at the outset. So we'd want people to undergo a serious program of training before actively participating. Ideally, we'd like to see all representatives live on a "campus" and spend a lot of time with each other. No flying back to the home district Thursday night and returning Tuesday morning. The US House of representatives was in session only 111 days in 2016. I guess its members have better things to do than run the government [Bresnahan 2015].

We'd recommend that representations be given just a single term, say 4 years, to head off the "revolving door" of current politics.. We might even consider giving each representative a pension they could live on afterwards, just to decrease the temptation for corruption while in office. If, after leaving office, they are found to be corrupt, such retirement would be withdrawn and they'd go to jail. Terms are staggered so that only 25% of the legislature changes in a given year. We'd also want to build in some way to encourage expertise and organizational memory, perhaps by having variable-length terms or keeping past participants involved in an advisory role.

Can it work?

In the longer term, the technical and economic revolution of Makerism we're advocating will render much of what today's government does, obsolete. By ending physical scarcity and designing person-scale systems, we can eliminate much of the infrastructure that government now needs to coordinate. Hyperlocal control decreases the need for most big-scale cooperation. A household can handle most of the coordination within itself. Neighborhood-level can do much of the rest. (see the chapter *A day in the post-scarcity life* [Ch. 18].)

The new kind of government that this chapter proposes is untested. We can't claim that it has been shown to work. Today's legal, economic and education systems lock in the status quo and will "self-correct" to prevent change.

An ambitious plan? Sure. Tough to implement? Of course. Necessary for the survival of civilization? You be the judge.

Recap

In this chapter we provide a new architecture for government. Key points are:

- Rethink the idea of representation in government. Now, representatives are supposed to be elected because of their positions on issues, thereby tying people's fates strongly to particular ideas. This makes it difficult for them to look at issues objectively, and difficult to cooperate with those of differing viewpoints.

- Remove external incentives for unjustified advocacy: election/ re-election; bribery (= "lobbying").

- Choose representatives at random, like jury duty. No gerrymandered districts, entrenched incumbents, etc.

- Explicitly educate representatives, government officials, and the public about cooperative processes and the scientific method. Make debate and decision-making transparent.

- Allow every citizen to participate in government by a public online forum that lets citizen propose ideas or offer opinions. Successive layers, moderated by professional, disinterested facilitators, summarize lower levels of discussion and elevate innovative ideas.

- Build in explicit mechanisms for cooperation and conflict resolution. Most of the time in government should be devoted to joint fact-finding and construction of win-win proposals. For some of the techniques, see the discussion of consensus process in the *Some of Us* chapter [Ch. 8] and its references.

- Develop technology that supports cooperative problem-solving processes. We will explore some in the chapter, *Tools for Reasonocracy* [Ch. 27].

Chapter 26
Some days in the life of a Reasonocrat

Aisha's head was still feeling a bit fuzzy from last night's bachelorette party, as she was staring at her fruit wall trying to decide whether to pick a mango or a papaya for breakfast. The wall grows enough fruit to supply a piece for each of the three household members every day, forever. It's timed so that 3 ripen each day.

She glanced at email on her phone. The first subject line read,

> Congratulations, Aisha, you have just won the United States…

She didn't even need to look at what it was she supposedly "won". Obviously, the message was spam. Funny, her filter was usually pretty good at eliminating spam. Odd, also, was that the message bore an icon indicating an encrypted and verified digital signature. Why would a spam message have that? She was about to mark it spam so that she wouldn't get any more of those messages, when curiosity got the better of her and she opened it.

"Oh, my God!"

From: Tony Hamilton <tony@uscongress.gov> (Verified Signature)

Subject: Congratulations, Aisha, you have just won the United States Reasonocracy Lottery

Official US Government Business

Dear Ms. Hernandez:

Greetings from the Congress of the United States. You've been randomly selected as a candidate for the US Congress. Details at

   www.becoming_a_reasonocrat.gov

Please phone me ASAP.

Tony Hamilton

Director of Recruiting and New Member Education
United States Congress
The Capitol, Washington, DC 20001
202 555 1212
Public key: hj3jhk24348787sdf0239485554234585b

You may be wondering, Dear 2018 Reader, just what is a *Reasonocrat*? Let's look it up in the dictionary.

Reasonocrat, n. A person who represents the general population in deliberations by the government about how to solve problems. So called, because the processes are designed to try to reach consensus by using reason instead of power, inspired by the social processes of science.

Antonyms:

Bureaucrat, n. A person who implements government policy via fixed routine without exercising intelligent judgment.

Aristocrat, n.: A person who gains great power and/or wealth via heredity or conquest, and makes decisions for those beneath them in the power hierarchy.

Democrat, n.: A person who believes that voting is the best way for a group to make a decision.

The "antonyms" share this characteristic: they all resist change from the status quo. This despite overwhelming evidence that bureaucratic, aristocratic, and yes, even democratic processes cause unnecessary war, poverty, and unhappiness of the governed.

Aisha thought she should visit the web page first. It requested that she login, then said:

> Congrats Aisha, you're one of a select few that are invited to apply for the position of Reasonocrat in the US Government starting next year. The requirements are so few that most people over 18 meet them, and you wouldn't have received the invitation if we didn't already expect that you met them.

Hmm, I've got a three year old to raise, we've just started planning a new house, and well, now is not a great time for this "opportunity". She dialed the phone number:

Tony: Hi Aisha, glad you called.

Aisha: (After pause.) Ah, I was expecting 5 levels of voice menus.

Tony: Yeah, that was the old bureaucracy. Turns out, voice menus wasted everybody's time including the organizations that deployed them. Besides there's only 4 people I give this number to. Managing your application is very high priority.

Aisha: What is your job again?

Tony: I help people overcome the shock of winning the lottery, so to speak. We still call it a "lottery", but it's not like that old game that was used to scam people back when we had Capitalism. Sad to think that people were so desperate then, that they pinned hopes on winning money. Also that they were so uneducated that they fell for it. My fifth-grader's Probability class would know better.

Aisha: I'm not sure I want to win.

Tony: My job is to help you decide. You can call me any time with a question about anything. OK, I don't know everything but as a former Reasonocrat myself, I pretty much know what's in store.

Aisha: How much do I get paid?

Tony: The same as me and what you make now, nothing. But the digs are pretty comfortable. Your own apartment, there's a swimming pool, theater, and food of pretty much any nationality you'd want to eat.

Aisha: Can I bring my family?

Tony: The little one, sure. James? Let's see you've been living together over a year, right? We should be able to get him in too.

In a month Aisha, Trish and James are on a plane to Kansas, chosen for the government because its literally the center of the country and no big cities. Tony greets them at the airport and they drive 45 minutes to the Reasonocracy Center. Though it doesn't grant any degrees, everyone calls it "The University".

Aisha: Why do you call it that?

Tony: You're here for 4 years and most of your time is spent learning. In fact, that's all you do as a freshman. Best professors in the world, small class size, very interactive.

Aisha can see the website photos were accurate. She already felt familiar with the physical layout. She and the other applicants are lead to the main auditorium. After an overview presentation, there's time for questions.

Aisha: I can see what the place is and can imagine the process, but why would I *want* to come here?

Prof: Well, I can't get inside your head but I can tell you that only 1% don't make it through the 4 years and nobody at the end wants to leave. Ask Tony why he wanted to continue being involved. You'll learn a lot, you'll make a real difference, and you'll have more respect than you can use from the whole country. Its nothing like being a lawyer or politician.

In a different venue, James was assured that there were a lot of opportunities here for him too, starting with an extensive maker space. Learning, interacting, helping out with the tasks of making a village run smoothly. That village would be raising their daughter too, in about the most cooperative and empathetic way possible.

Aisha: Will you still love me if I want to do it?

James: I'm not sure I would if you didn't!

Two Years Later ...

Aisha is on the "Peace Advocates" committee. After the civil unrest of 2025, "The Police" became just the name of a 1980s rock band. The women and men in blue changed their name to "Peace Advocates" to reflect a whole new philosophy on policing. With scarcity solved, jobs are, what would have been called, "volunteer". Even so, prestigious jobs like Peace Advocate are in demand and have stringent requirements. The overall need for police is down thanks to scarcity mitigation. However, some aggressive people, and mentally ill people still commit crimes. We listen in on a Peace Advocates committee meeting.

Member 1:   I've just been sent a final draft of the Oakland Peace Advocates rules of engagement with citizens. They're proposing these be incorporated into the training program for national Peace Advocate certification.

From "The University" education, Lesson #3: Before you take a position on an issue, the first step is to identify the relevant criteria for making the decision.

Aisha: What does it say about use of lethal force?

Member 1:   Peace Advocates aren't allowed to carry lethal weapons of course, but they do suggest sticky foam and a new chemical that blinds you for about 15 minutes.

Member 2:   A guy in my old neighborhood got blinded by that stuff a few years back, but he was still able to do a lot of damaging with the machine gun he was holding.

Lesson #7 Back up rationale with reality.

Member 1:   That's where the sticky foam comes in. Ok its not perfect but a lot better than the old days of shoot first and ask questions later.

Lesson #9 An ounce of prevention is worth a pound of cure.

Aisha: This is just the tech stuff. What about the kinds of events that trigger stopping citizens in the first place?

Member 1:   Its kinda complicated. There's lots of "indicators" but at the end of the day, it's still a judgement call.

Lesson #12: Expand the solution space.

Member 3:   In England they've got a different set of rules and, according to statistics, its working out pretty well. Why don't we just adopt those?

Member 2:   Do any of these solutions train officers in Psychology? You know, what motivates people, what makes them angry?

Lesson #14: Invention is always on the table.

Member 1:   Hmm... the Oakland proposal doesn't and I'm pretty sure England's process is minimal in that area. But that's an important idea. Let's have our recommendation include some psych training.

A couple hours later...

Lesson #21 Rationale is important but hard to manage...

Aisha: OK we've captured a lot of good rationale in our Justify discussion about "rules of engagement". What seems apparent to me is that we've got a ton of good ideas, but we differ on how effective those ideas are in actual practice.

Lesson #15 Recognize ignorance.

Member 1:   As fond of the Oakland policy as I am, I have to admit that they don't have enough data for us to be confident about recommending their proposal for National deployment. Heck, this is going to determine life and death.

Lesson #2 Embrace the scientific process

Aisha: While lives are being lost, we must not dilly-dally. On the other hand, I don't want to make a hasty judgement here. Let's play the "Scientific Experiment" card. We've got one just started in Oakland on their plan. That covers a tough inner city. We need a suburban and a rural setting, and we also need to contrast the Oakland plan with England's.

Lesson #18 Haste makes waste but time is of the essence. Its a balance.

Member 4:   England isn't America.

Member 2:   America isn't America, but so what? Let's catalog the differences and learn as much as we can from the commonalities.

Aisha: We're going to need some big data analysis here. Let's ask for a couple suburbs and a couple rural areas to implement the Oakland plan, then rope in some statisticians to carefully get the raw data from them *and* from England.

Lesson #2b: Generate and test, experiment and analyze.

Member 3:   And don't forget that we'll need the machine learning guys to help generalize from the megabytes they'll collect.

Aisha: OK, we'll let the Logistics Committee hammer out the details, then revisit this issue in a year to see if we're getting a clear direction from the machine learning guys. Remember we'll take their input with a grain of salt. All the reasoning we've captured here today and then some will be needed to help shape the best solution.

A few years later, another issue comes up …

Friend: The guy next to me has a zillion of those noisy little drones flying to his house every day. Since you're a former Reasonocrat, I though maybe you could put in a word with the powers that be and …

Aisha: Hold on. Former Reasonocrats have no revolving door to K Street like the old US Democracy. I'm just an ordinary citizen now.

Friend: So you can't make those pesky things illegal?

Aisha: *I* can't but *we* can give it a shot. Let's talk to our town's Environment Committee.

Two days later …

Env Committee member: We've had a noise complaint about the constant drones flying to your house.

Neighbor: Yeah, it bothers me, too. But here's the thing. I've invented
               a new kind of magnet that helps doctors locate tumors and
               now every hospital wants one.

Env Committee member: Hmm, I know this aerodynamics guy.

Later ...

Aerodynamics Guy: Turns out if you put bumps on propellers, à la
               whale flippers, it can increase aerodynamic efficiency and
               decrease noise. I know this drone guy...

Drone Guy: OK. I can 3D print some new propellers for the drone I'm
               working on. I'll upload them to thingiverse.com. Also, I'll
               enter the rationale for their design into the Environment
               Committee's citizen feedback site, so that the government
               will know that they have a possible solution when someone
               else complains about drone noise.

Neighbor: Since I want to be a *good* neighbor, I'll suggest to the
               hospitals that they start using the new, quieter drones.
               Speaking of quiet, it'll quiet down my neighbor who
               originally complained, to know that there's a solution on
               the way.

Chapter 27
Tools for Reasonocracy

We assert that reasoning is a better process for decision making that power. Let's put some stakes in the ground to get practical.

Goals

Our highest-level goal for government is the greatest good for the greatest number with minority rights. How might we compute that?

Good-for-each-individual x number-of-individuals

This is a simple metric, but often too simple. Suppose we have 10 individuals. Policy A changes the income of our richest individual from $100K per year to $200K per year but decreases the income of the other 9 from $50K per year to $45K per year. Is it a good policy? Total wealth gained is $100K—(9 * 5K) = $55K clearly a net positive.

But we are comparing different wealth individuals, so another way to analyze it is to add up the percentage gains: Our rich guy goes up in his wealth by 100% (doubling his income), whereas each of our 9 go down by 10%. 100-percent-points—90-percent-points = 10-percent-points, still a net gain but not as much of one. A 3rd way to analyze this is that 1 guy makes more and 9 make less, so if we balance our winner with 1 loser, our net is 8 losers. So counting "whole team score", policy A is a big win, but when counting individuals, its a big loss.

What does Capitalism do? Well neither really. Capitalism tends to make it easier for the rich to get richer and the rich make decisions (like tax rules) to benefit themselves. So Capitalism keeps score in a 4th way: If the rich (few) do well, that's good; everybody else doesn't matter much. Note that our 4th way happens to correspond to an overall higher "team score" in the above example but that's an accident in Capitalism, not a designed-for criteria. Had the poor people lost more than the rich guy gained, team score would be negative but Capitalism would still be for it.

Our example is not so hypothetical. Especially in the last decade, US GDP is up, wealth of the wealthiest 1% is up, but poor people's income is down. With 2 benefits, the status quo can claim its maximizing team wealth and back that up with overall GDP numbers, but the correlation between Policy A and team wealth is not causally linked.

Bhutan [Bhutan 2017] measures success by "Gross National Happiness". This is more complex than GDP, so its success is more complex to assess. But its a pretty compelling idea, with the caveat of how to manage the complexity.

The Devil's in the details

Gross measurements of dollars or happiness are crude tools. And assessment is only part of the story, as any opponent of "No Child Left Behind"'s educational testing policy can tell you. Measurement alone does not provide new solutions.

Negotiation

Some of our most powerful tools are conceptual. Such rather loose concepts can be packaged into processes that make them actionable. In *Good for You, Great for Me*, Larry Suskind [Susskind 2014], one of our favorite MIT professors. Teaches how to make negotiations more beneficial for both (conflicting) parties. An example is in order.

Many cities have laws limiting the number of liquor licenses they will grant. A restaurant must have one to sell alcohol. Our example city has distributed all the allotted licenses, yet 2 unfriendly business men each want one. A restaurant owner retires and the license reverts back to the city. The two business men each want the license.

Scenario 1: Each presents to the city council why they should be given the license. Joe hires a pro marketing firm that designs presentations and wins.

Scenario 2: Sam learns that Joe has hired a marketing team so he does the same thing. Then Joe hears about this and gives his team more money. Joe and Sam are now in a classic "arms race" where the real winners will be the arms merchants (in this case marketing firms). Eventually either Joe or Sam will win but the victory will be reduced by the marketing expenses.

Scenario 3: Joe learns Sam is also employing marketers and realizes they are in an arms race so he fires his marketing staff and uses his marketing money to bribe the councilmen. By switching "media" he hopes to avoid an arms race, but just creates another as Sam switches to the same tactic.

Scenario 4: Sam realizes it doesn't matter which battlefield the arms race is on (marketing or bribery), it still drains profits of both combatants. He convinces Joe to pay the other half of the fee of a Negotiation facilitator named Larry. Larry's strategy is:

1. Think of some new options that neither Joe or Sam could come up with on their own.

2. Help them collectively choose the one with the best outcome, trying for a win-win in a situation that originally seemed like it had to be a win-lose.

Scenario 5: The first new card Larry puts on the table is maybe we can convince the city council to split the original 1 full liquor license into two beer-and-wine-only licenses. Estimating from other businesses, they determine each would get about 40% of the revenue of the full liquor license.

Scenario 6: Larry doesn't like losing 20% of the revenue. Joe and Sam aren't happy about it either, yet it still sounds more attractive than a 50% chance of getting nothing. Next solution is to have Joe and Sam co-own one new restaurant, sharing costs and profits. Now each is getting 50% of the profit so "the team" isn't losing 20%. This is even better. Each spends 50% of what they would have to get the restaurant by themselves, and each gets 50% of the profit.

Scenario 6: In trying to seal the deal, Larry realizes that each is a little hesitant: they each are bothered by not getting that 100% of the profits. But they each still have half the capital to start a 2nd restaurant. So Larry proposes they form a "chain" and explore liquor licenses in nearby towns to also co-invest in. Each investor likes the increased diversity this new approach affords, making the total investment more stable. Deal Sealed.

What can we learn from this?

1. By getting creative, part of the wealth of a winner can compensate a loser. This technique is especially important in the "cut the baby in half" compromise situations.

2. By cooperating, an expensive battle can be avoided, benefiting both parties.

3. By "expanding the pie" a "divide down the middle" compromise doesn't have to mean that each party gets a smaller piece.

Negotiation is generally restricted to a few parties. But a government has millions of parties to appease and a myriad of diverse issues to resolve. We need some new tools.

Justify

To help people understand how their own ideas relate to the ideas of others, we wrote a program called *Justify*. Justify is a language for expressing rationale, and a "development environment" for organizing those ideas to help people model complex situations and make better decisions. Justify tries to encapsulate each idea or opinion in what it calls a *point*. Roughly, a point is what you might write in a single post in an online discussion. Justify guides people to declaring what kind of "point" they are making (one of a couple hundred different kinds) and where that point should go in a hierarchy of related points.

This architecture doesn't restrict *what* can be said, nor *when* something can be said, but by restricting *where* and *how* something is said within a hierarchy, it does help mold complex ideas into a coherent discussion. Justify also summarizes each point and its sub-points with an automatically created *assessment*. This helps humans understand the implications of their reasoning at each level, and helps the program make even higher-level assessments. Navigating a complex hierarchy is facilitated by the user seeing, at each level, an assessment of lower-level points [Fry 2013].

We propose Justify as a standard venue for government's decision making process. Justify's assessments don't depend on *who* said something (like power-based governments), but rather on *what* was said and especially *why* it was said. Because points are organized semantically, not chronologically, the order that statements are made in is unimportant, unlike the debates and ordinary discussions of democracy and consensus.

Above we outlined how a skilled negotiator can find opportunities for value for two conflicting parties and reach agreement. Sadly, skilled negotiators like Larry Suskind are rare. Can we make a tool that might lead participants to a similarly positive outcome?

Language

We can steal a winning idea from programming: To solve a problem, first define a language to describe problems, then use that language to describe *particular* problems and their solutions. Programmers use general purpose programming languages, like Python, and JavaScript, to model the underlying issues of their applications and to create new facilities to work with those issues. But programming is complex, so special programs called "development environments" have been created to facilitate working with programming languages to create new programs.

Part of Justify is a new language to describe rationale for making decisions. Like a general purpose programming language, it is not targeted at any particular kind of decision. It tries hard to be "general purpose" enough for any kind of decision. A small group discussion can use Justify's point type language to express the role that each of their contributions plays in the overall argument. Justify is implemented in the general purpose programming languages of Python and JavaScript, and runs on the web, so anyone can use it. http://justify-app.appspot.com/.

A tale of two governments

Let's walk though a concrete example with two different processes. Our example is in deciding the material for a new bridge. The bridge will provide transport for a million cars a year and hopefully last decades. Millions of dollars are at stake. The transportation committee has narrowed down our viable options to two: carbon fiber and (we said it was a concrete example) concrete. Let's listen in.

Power-based government scenario

In the power-based discussion, we hear lots of reasonable sounding rationale. It comes from the companies pushing product, the legislators and the presumably neutral university experts. This material has a proven track record. That material is stronger. This one's cheaper, that one is corrosion resistant. This new formula overcomes the previous defects, and so on. Some of it is accurate, some of it isn't.

But it really doesn't matter because, in conventional politics, it's not a reason-based decision making process. Behind the scenes dollars are passed, promises made and deals are sealed. In a complex situation, you can rationalize anything. The side with the most under the table wins. Yes they need to have the pedigreed experts, the "conservative" spreadsheet, the "makes sense" sound bite. But the other side has all that too.

What it comes down to is the most cash (or the equivalent) in the right pocket at the right time. Power based governments back the status quo. Concrete wins over carbon fiber. Innovation occasionally wins, but only occasionally.

Reason-based government scenario

Superficially, a reason-based process can look very similar. Expert testimony, rational, rebuttals, etc. The difference is that what is said *actually* matters. There isn't a back room where the deal is sealed.

Because a reason-based process would allow for *learning*, it's likely that we would come up with possibilities beyond the original two. Our final design gets the benefits of carbon fiber where its needed most, and the benefits of concrete where that is most appropriate. We don't end up with any big winners. Manufacturers are compensated fairly with a profit that allows them to do their best work, but not retire. Citizens get a fair deal for their tax dollars, not something that sounds too good to be true. Legislators had to work hard to hammer out the deal, but didn't have to break laws, wallets or hearts to do it.

With these themes in mind, you're prepared for the next level of detail. We frame our issue with a question.

Assessment ←	Type		Point Title		Info
▶ unfinished ←	?		Is concrete better than carbon fiber?: *most*		

Our Justify point has a title representing the question. The pink "?" represents the point type. The "Assessment" is "unfinished because there is no rationale for answering the question, indicating that we need more detail. Our first rational is one in favor of concrete:

◢ pro 1   ⬅ ? Is concrete better than carbon fiber?*: most*
▷ pro 1   ⬅ 👍 Yes its cheaper per pound.

We have added a "pro" point that assesses to itself and, because its the only reationale we have, becomes the assessment of our question indicating that concrete is best. However, the carbon avocates rebut:

◢ unfinished ⬅ ? Is concrete better than carbon fiber?*: most*
　◢ refuted   ⬅ 👍 Yes its cheaper per pound.
　　▷ con 1   ⬅ 🔖 but not cheaper per strength

Our new con point assesses to itself, and causes the assessment of "Yes its cheaper per pound." to become "refuted". Because that 2nd point is refuted, it no longer counts for our top level assessment, which reverts to "unfinished". An open minded participant asks an honest question.

◢ unfinished ⬅ ? Is concrete better than carbon fiber?*: most*
　◢ unfinished ⬅ 👍 Yes its cheaper per pound.
　　◢ unfinished ⬅ 🔖 but not cheaper per strength
　　　▷ unfinished ⬅ ? Why does this matter?

Because this question has no possible answers, it is "unfinished and that "unfinished" propagates all the way to the top. Letting us know we have work to do.

◢ unfinished ⬅ ? Is concrete better than carbon fiber?*: most*
　◢ refuted   ⬅ 👍 Yes its cheaper per pound.
　　◢ con 1   ⬅ 🔖 but not cheaper per strength
　　　◢ folder(1) ⬅ ? Why does this matter?
　　　　▷ carbon is...⬅ 💡 carbon is 500x stronger, only 60x $

Giving our 2nd question an answer makes it no longer unfinished, and restores the refutation of "Yes its cheaper" 3 levels up. WE have more information but still no resolution to our top question.

◢ con 1   ⬅ ? Is concrete better than carbon fiber?: *most*
  ◢ refuted   ⬅ 👍 Yes its cheaper per pound.
    ◢ con 1   ⬅ 👎 but not cheaper per strength
     ◢ folder(1)   ⬅ ? Why does this matter?
      ▶ carbon is…⬅ 💡 carbon is 500x stronger, only 60x $
▶ con 1   ⬅ 👎 Carbon is smaller and thus prettier

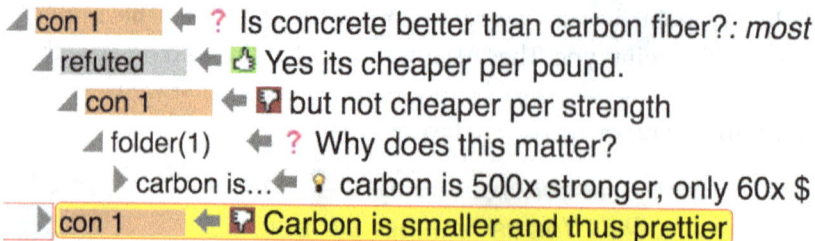

Our last point is not addressing the "cheaper" aspect, but going for a new rationale immediately under our top level question. Now our top level question has one con point beneath it and no other unrefuted points, so its assessment is "con", i.e. concrete is not better than carbon fiber.

Complexity

In a 6 point discussion, humans can keep track of the rationale and what it "add up" to. But imagine the hundreds of points in a real debate about safety, durability, capacity all with their own cost and aesthetic considerations. When humans are faced with overwhelming complexity, they tend to throw out rationale and "go with their gut". In a complex world, such emotional strategies cause sub-optimal choices.

Justify comes up, not simply with the top level answer, but answers to embedded questions and the implications of rationale at each level. Its reasoning is transparent for all to see. If a new situation arises, say a new high strength concrete is developed, we can revisit our discussion without having to start all over. The information about carbon and aesthetics is still valid. Other points can be placed in the new context, **not** by changing those points, but by adding new rationale. Thus our historical reasoning is preserved, yet refined to a more accurate picture of the new reality. This allows new entrants to our discussion to both understand what preceded them, and add to it in a way that builds greater understanding.

Justify enables a user to not have to wade through volumes of arguments that they don't care about, yet facilitates diving down into the details of those points they do care about. Those triangles to the left of each point allows shrinking or expanding the points under any point. This permits not just a superficial high level view, or an "all details" view, but rather, detail where you want it and not where you don't, all a few mouse clicks away.

Can we afford government by reason?

Above we've presented two different techniques that can improve upon "emote and vote". Each requires more effort than our typically simplistic voting schemes. (Or does it? The US 2016 presidential election cost billions of dollars.) A more reasoned approach, even if it made government just a few percent better, would surely pay for itself.

An innovation mind-set would encourage inventing new techniques for improving the reasoning capacity of decision makers. The development and deployment of these advances needs education, the will to improve our world, and persistence in the face of a repressive status quo. Expensive to be sure. But, we believe, cheaper than the alternatives.

Part 5
How can we
get along in…?

Chapter 28
Constructionism: Education for Makerism

For those of you who came to this book expecting us to present a panacea for solving all the world's problems, we're happy to report that we do indeed have one: *Education*.

There's hardly a problem that we talk about in this book that wouldn't be helped enormously by improvements in education. Our major goal in writing this book is to convince people to be more cooperative. Educated people tend to be more cooperative, and cooperative people tend to be more educated.

But we don't want to rely merely on correlations. We think a good chunk of education ought to be devoted explicitly to tools and techniques to help people to cooperate with one another more effectively. Besides solving adversity, and increasing productivity, we can learn to solve some of our most persistent problems by being more cooperative.

Poverty? In the short term, education helps people get better jobs. In the long-term, we believe that education will help people advance to the point where poverty (and jobs!) are eliminated.

Oppression? Educated societies are less susceptible to exploitation by messianic religions and dictators.

Crime? Crime rates go down as education levels increase.

Health? Educated people know more about their bodies and tend to have healthier lifestyles.

War? The USA spends more money on war than any other country, despite the fact that our citizens are considered to be highly educated. This apparently contradicts the "panacea of education" described in the opening of this chapter. First, wars are caused by leaders, not by average citizens. Now, it's true that citizens elect the leaders, so we're not letting the citizens completely off the hook. And the astute will observe that America's leaders often graduate from top universities. With innovative research and education, we can design and deploy tools for governing that are more rational than causing war (See *Reasonocracy* [Ch. 25]).

Instructionism: The factory model of education

We argue that our present justice, political and economic systems are based on principles that are rapidly becoming obsolete. So too will we argue that the structure of most of today's education systems are also organized along the principles of the Industrial Age. Far too many American schools implement what education reformers call *Instructionism*, or the *factory model of education*.

A school is basically a factory for producing *human robots*—obedient factory workers that can perform repetitive tasks commanded by the power hierarchy. Bureaucracies are basically factories for human processing of information.

If there's a task to be done that can be done "robotically", let's get an *actual robot* to do it, not a human. Makerism and disintermediation replace bureaucracies and hierarchies with do-it-yourself production and cooperative organization. Why do we need schools to educate people to be robots?

Today's schools aim to impart a minimal set of reading, writing, and math skills necessary for following orders in the workplace. Only incidentally, if at all, are schools genuinely concerned with helping students lead happy, productive, and fulfilling lives. Students toil at the assembly line of rigid classrooms and standardized curricula, with quality control of the product enforced by endless testing and grading.

Note that *many* teachers are personally unhappy with Instructionism, but their more inclusive and diverse teaching styles are hampered by controlling Common Core curricula and other test-happy bureaucratic rules.

If the Industrial Age, and its factory schools, are so lacking, what can better education look like in the Makerism age which is now upon us?

Constructionism

There are many ideas put forward by education reformers, and implemented on modest scales, that have merit. These go by many names: *alternative education; student-centered learning; exploratory learning; project-based/experiential/hands-on learning; Montessori method; Constructionism.*

This last term, *Constructionism* [Papert 93], refers to the educational philosophy of Seymour Papert, based on Piaget's theory of child development. It advocates using computers to provide *microworlds* that allow students to explore powerful ideas, acting like scientists in formulating and testing theories. It led to the programming language *Logo*, and its latest incarnation, *Scratch*, used by millions of children. Author Lieberman was a member of Papert's original research group at MIT. He was also a student and teacher in MIT's Experimental Study Group (http://esg.mit.edu), another stronghold of the Constructionist education philosophy.

There is a vast literature on reform of education, dating at least back to the 19th century, when the factory model for education first appeared (and, coincidentally, when factories first appeared). Maria Montessori [Motessori 1969], John Dewey, Ivan Illich, Paolo Friere and others wrote about how students should be active participants in their own learning, rather than passive recipients of knowledge. In 1921, A.S. Neill founded the Summerhill School [Neill 1960], which put these principles into practice and his book remains a classic. We also recommend the "project based" (as opposed to test based) curricula of San Diego's "High Tech High" as exemplified in the documentary *Most Likely to Succeed* [Whitely 2015].

The role of schools and teachers should be to help students *learn how to learn*, and to provide a community of peers to help each other learn. Unfortunately our current practices and data-driven school systems often fall short of this objective.

Cooperation and competition in education

Educational philosopher Alfie Kohn thinks that the primary subject taught in US public schools isn't math, English, or chemistry: it's *how to compete*. Grades, sports, and vying for teacher praise, all pit students against one another. No wonder adults have such a fondness of war.

We argue that technological and social changes are increasingly favoring cooperation over competition. In his book, *No Contest: The Case Against Competition* [Kohn 1986], Kohn systematically lays out the advantages of cooperation over competition in education.

Kohn also shows how competition among students, teachers, and schools is antithetical to achieving educational goals. The disastrous No Child Left Behind and Common Core movements emphasize standardized curriculum, competition, and testing. They provoke outrage and despair amongst students and concerned adults. It'll be years before we completely recover from them. Kohn presents countless ways in which education can be made more cooperative.

Kohn also puts his finger on one of the most important issues in education: student motivation, in his book, *Punished by Rewards* [Kohn 1993]. As soon as you finish this book, the two Kohn books should be next on your reading list.

Motivated students learn, to the best of their ability; students who are not motivated do not. But there are two kinds of motivation: *intrinsic motivation* and *extrinsic motivation*. We explored the differences between them, and the fact that extrinsic motivation inhibits intrinsic motivation, in *Intrinsic and Extrinsic Motivation* [Ch. 10].

Competition is extrinsic motivation. This kind of motivation, like educational games and grades, may be useful in the short term to spur students to become exposed to a topic they might not otherwise explore. However it is only intrinsic motivation that ignites the passion that animates true learning in the long-term.

Therein lies the real difference between the two visions of education. The factory model requires that motivation for learning be imposed externally. It completely ignores the internal motivation of the learner. It is completely oblivious to the interests and idiosyncrasies of each person.

The standardized curriculum tells you what to learn next, not allowing for variation. It is oblivious to learning for its own sake, or if the student is lost. Rewards that the student feels from learning must come from success in passing tests and receiving high grades. Because the competition for grades is a zero-sum game, there will be few winners and many losers. Where does that leave the losers?

The Constructionist vision of education, on the other hand, emphasizes the intrinsic motivation of the student. It encourages students to follow their own interests, express their own unique personality, and share their interests with others. It encourages each student to become a creative problem solver—formulating theories, learning from experience, sharing their passion. It also seeks to cultivate emotional and social intelligence in students.

The role of the teacher is less like a factory foreman, and more like a personal trainer—encouraging the student, nudging along the student's intrinsic motivation, providing training that helps acquire needed skills, and helping the student get unstuck when they encounter difficulties.

Meta-knowledge

The essence of the Constructionist model of education is basically to turn students into Makers. What do you need to learn to be a successful Maker?

First, we'll tell you what you *don't* need to know, because simply piling on more things to learn doesn't scale. There is not much point in being able to memorize large collections of obscure factual information when you can simply look it up on the Web. There is not much point in being able to flawlessly execute procedures that you could program a computer or robot to do. The stark reality is that you won't look information up, or program it, unless you are curious and interested.

Curiosity, passionate interest, and *resourcefulness,* are examples of skills and traits that will become increasingly important. When production of hardware and software becomes increasingly automated, it's the creativity and good judgment about what to produce that's essential. Troubleshooting and debugging skills are crucial, because, as we know, when you use technology, things don't always go right the first time.

The biggest irony about our education system is that its goal is to impart knowledge, yet it doesn't impart much knowledge about knowledge itself. So the most important kind of knowledge might be *meta-knowledge*. Students need to learn not just how to learn, but how *they* learn best.

Often, people say they have "intuition", when they think they know something, but they have no idea *how* they know it. Intuition is not a thing, it's a lack of something. That something is the knowledge about how you know that something is true, i.e. its rationale. This is an important kind of meta-knowledge.

If people can introspect and articulate about the reasons for something, they don't have to rely on intuition—they can access their meta-knowledge.

Understanding yourself is perhaps the most important meta-knowledge you can obtain. Even meta-knowledge is best utilized when it's in the context of the learner knowing about how they learn best. We all have cognitive biases that cloud our judgement. We can compensate, at least to some extent, if we understand cognitive biases. A great article on ignorance is [Dunning 2014]. If you're confused about why Americans vote against their own self-interest, read this article. And, last but not least, you sometimes need to be able to figure out what it is that you *don't* know.

Complexity

Many of our largest problems remain unsolved simply because they're complex. So we should go meta on complexity, and study complexity itself. So what academic field best gives us tools for complexity?

Well, we're programmers. Our answer is that computer science has developed some extremely powerful, and under-appreciated, tools for managing complexity. Chief among them is the idea of *integrated development environments (IDE)* for authoring programming languages.

Why are these so powerful? In problem solving, certain ideas and techniques tend to occur again and again. If you encapsulate these ideas in a language, naming them with words, you vastly increase your ability to solve problems by composing phrases. The idea of a development environment is software that helps you manage this use of language.

It helps you compose words and phrases, test them, and fix problems when the occur. For example, you could think of Microsoft Word as an interactive development environment for English authors.

The power of a formal language with an IDE shouldn't be limited to programming and English. Even the large-scale problems we discuss in this book, such as the economy and government, could benefit from this approach. We develop these ideas further in the chapter *Software Makerism* [Ch. 16]. We present a specific design for a language and environment for decision-making in government in the chapter *Tools for Reasonocracy* [Ch. 27].

Leaning about life skills

Schools fail to teach life skills. Some of the most important aspects of contemporary human life go almost completely unmentioned in school curricula.

The details of getting a job, buying a car, eating and exercising properly, moving to another city, maintaining romantic relationships, maintaining a home, or what to do when a cop pulls you over, receive almost no formal instruction in schools. Go find those subjects in the Common Core curriculum.

If your parents or peers didn't teach you properly about such things, you'll probably fumble through these processes on your own, and are likely to make serious mistakes that will drastically affect your quality of life. Universal education in life skills would almost certainly be cheaper and more humane than trying to mitigate the consequences of people's trial-and-error disasters.

Self-Sufficient U.

One possible way to create a community where life skills and education are integrated might be a project to create a "Self-Sufficient U." Can we provide the equivalent of a college education, without the $60K/year costs?

The school can grow its own food (aeroponics for fruits and vegetables, aquaponics for fish and seaweed, chickens for eggs and meat), build its own buildings (bamboo is strong and fast growing), generate its own energy (solar and wind) and handle most of its own health care (smart phones with peripherals are amazingly capable).

Such a school wouldn't have an Astronomy or Russian Literature Department, nor a football team, but agriculture, architecture, energy and medicine will be front and center. In addition to teaching life-skills, there will be classes in efficient business management (aka: cooperation), justice, government and, of course, education, will fit right in with this practical agenda. The best way to learn something is to live it, just like learning a foreign language is easier, when you're living in the country that speaks that language. The more experienced students can carry much of the teaching load, especially if mentored by seasoned professors. Students perform administration tasks including admissions. How better to learn about management or hiring? The education of "running your own show" is powerful and practical. It's also a motivating platform for research into self-sufficiency, which will become essential as Capitalism crumbles.

Can this be done? An agricultural high school in Paraguay has developed a model for self-sufficient schools [School-in-a-Box 2017]]. Tuskegee University in Alabama was founded in 1881 with very little money and a program wherein students maintained farms and built campus buildings to cover costs. The Israeli Kibbutz had success in the 1960's. While TU and Kibbutzim may have diluted their initial focus, with tech advances, self-sufficiency is easier now and will get even easier.

Online education

Technology is providing even more educational opportunities. Online courses, and resources like MIT's Open Courseware, are available for almost every academic subject. TED talks provide insight from the world's great thinkers. Automated translation will make the world's knowledge available to people who speak minority languages in remote places, and help them share their unique perspective with the world. Peer-to-peer learning will help alleviate teacher shortages. Intelligent tutoring systems will make online learning more effective and personalized. And, of course, online learning holds the promise of reducing costs and making education more accessible to all.

But that's a worse education, right? According to a report from the US Dept. of Education "The meta-analysis found that, on average, students in online learning conditions performed modestly better than those receiving face-to-face instruction." Consortia of universities are putting their courses on line and charging very little if anything for them. EdX and Coursera, just to name two systems, attempt to capture the best courses from Stanford, Harvard, MIT and a host of other top universities.

Making curricula

Constructionism has, at its core, a philosophy of allow students to learn by constructing things. By becoming a maker. Going meta, one of the most instructive things to make is a curriculum.

One of the best ways to learn something is to be put into the position of teaching it. You're motivated to do a good job because you don't want to be embarrassed in front of the students, This also forces you to take a student's eye view of the information you'll be presenting, causing the kind of reflection that's likely to improve the educational presentation of the subject matter.

But are there enough students to go around to make everyone a teacher? Yes, if there's just one student you teach: yourself.

Today billions of people teach themselves things when they use a web search engine to find out things they don't know. A quick way to find what you're after is what we call *Search By Design*. First you design (to a low resolution) the solution you're looking for. Now that you know its parameters (roughly what it looks like) you can form more targeted queries that return more relevant results. Plus, because you know what it should look like, you have a much better chance of recognizing it when it shows up in your query's results.

This works when searching for isolated things, but isn't so effective for learning an entire subject. That's better done with *Learn By Design*. You design the curriculum that you *think* would best meet your needs. Now you can search for it with a greater chance of success. Or you can *Search by Design* for the various pieces of your curricula. If you have a hard time getting what you seek, its an opportunity to help others by compiling what you do find, combined with your own design, into a new web page that will expand useful curricula for all.

Education: The real panacea

It's fashionably cynical to greet starry-eyed innovators like us with the warning: There are no panaceas. No silver bullets.

But if we set out to solve all the world's problems, improving education is about as close as you can get to a silver bullet. A *palladium* bullet, maybe (Silver is Element 47 in the Periodic Table. Palladium (which is even more valuable!) is Element 46.).

Chapter 29
Transportation

Transportation is the movement of people and goods from one place to another. It is the quintessential infrastructure. Transportation is vital to trade and to civilization.

Problems

Problems in transportation are endemic to most big cities, one of the principal disadvantages of an urban lifestyle. Traffic jams slow urban car, truck and bus travel to slower than a bicycle. The US kills 30K people per year in car accidents and hundreds of thousands are injured. If you include the health effects of air pollution and psychological stressors like "road rage", pretty much everyone is adversely affected. [Inrix 2016] reports some statistics from their study of 2013's traffic jams on France, Germany, UK and US:

- $200B wasted in traffic overall
- Average cost to a driver: $1740
- Average time wasted in congestion: 111 hours.

Oil

Oil is used mostly for transportation because it packs a lot of power per weight and volume (significantly more than state-of-the-art batteries). But its political consequences are high. [Klare 2005] articulates that most places oil is produced (Iraq, Iran, Nigeria, Venezuela, Russia, Saudi Arabia, USA, Canada) it causes big problems (pollution, war, propping up dictators). Since Klare's book, America has reduced its dependency on foreign oil, but at the high externality costs of fracking. The XL-pipeline is looming as yet another attack on the ecosphere.

Land

Land use by cars is so extensive that it causes there to be greater distances between the places where you want to go. From [Litman 2014] we learn: "in automobile-dependent communities with road and parking supply sufficient to keep traffic congestion to the level typical in U.S. cities, plus parking spaces at most destinations, a city must devote between 2,000 and 4,000 square feet (200-400 square meters) of land to roads and off-street parking *per automobile*. This exceeds the amount of land devoted to housing per capita for moderate to high development densities". Yikes!

Land has three variables guiding its price: location, location, and location. But what do we really get by being in a location? Most of it is "proximity to somewhere we go frequently". We care about proximity primarily because closer places take less time to travel to. But with faster transport, the disadvantage of being further away isn't as great. We could, for instance, live further from the city on cheaper land, and still have a short (time-wise) commute with a good transit system.

Cost

From [DOT 2015] the US Department of Transportation budget for 2015 is $90B. The preamble to this document quotes Obama: "We'll need Congress to protect more than three million jobs by finishing transportation and waterways bills this summer." Then in the very next sentence: "But I will act on my own to slash bureaucracy". Hint to the president: "bureaucracy is jobs". You can't simultaneously protect and cut jobs.

From our experience with federal and state departments of transportation, they have little to do with transportation. They are about jobs. Whether those jobs produce better transportation infrastructure is irrelevant to senior policy makers. What is relevant are votes for re-election or just maintaining the jobs within the department. Efficient allocation of resources is good.

Transportation projects are the classic pork-barrel projects and pork is not efficient. (Note to presidents, governors, secretaries: please prove us wrong. Read this chapter, call us, and we'll help you construct an efficient allocation of transportation dollars. Then we'll revise this chapter. Note to readers: If you're reading an up-to-date version of this chapter, the politicians are still not interested in improving transportation.)

Direct costs to a car owner are about $9K per year including depreciation, fuel, maintenance, insurance and taxes [Peterson 2014]. But that leaves out tolls, parking, having the land for a driveway, snow removal, health costs (including accidents and asthma), and the Middle East wars caused, in no small part, by demand for oil. For a typical car owner, their car is more expensive than health care, food, education and other categories. We also spend an awful lot of time commuting.

Solutions

Reduce transportation

Like solutions to the high costs of other infrastructure systems, our first strategy is to reduce the need for transportation by fulfilling its functionality in cheaper ways with fewer detrimental externalities.

Reducing people transportation

3D/holographic teleconferencing is coming. It *can* have visuals nearly as good as reality, and audio better than reality. So going into the office, or flying for meeting people will become unnecessarily wasteful.

By having home body sensors, much of the movement of people for health reasons can also be eliminated. Smart phones with attachments already allow people to take many measurements of themselves. More are coming. The phone can transmit these measurements to a doctor or a program in the cloud and appropriate instructions can be transmitted back. This is, after all, information transportation. Some other material goods needed for cures (braces, crutches, prosthetics) can now be printed at home. In the future, medications will also be able to be printed, saving trips to the pharmacy.

We can save many trips that people now do in order to get something by just moving the objects and a smaller vehicle. See Efficient Goods Transportation below.

Reducing transportation of goods

If you make (including grow) most of the material things that you consume, transporting them from the store to your house goes away, as does the good's transportation to the store. You will still need some raw materials, but with parametric designs, you can take greater advantage of local materials and reduce their transportation costs too. When you're making your own stuff, you can make just what you need, when you need it, so the inefficiencies of overproduction and spoilage also go way down. Packaging for shipment is eliminated, thus getting rid of yet more paper production (and foam peanuts). All that packaging is normally thrown out, but without it, our waste stream is *way* down. By composting bio-waste, and recycling the objects you print back into your printer to make other objects, there is very little waste left, eliminating trash trucks, their labor, oil use and pollution along with landfills and waste incineration.

Oil and gas are now transported in trucks and pipes. By getting more efficient buildings and vehicles, we reduce the need to transport oil and gas. By converting to electric appliances, generating the electricity with solar panels and storing it in batteries (all 3D printed and used locally of course) we eliminate the transportation of energy. That in turn eliminates the pipes, utility poles, and their wires, reducing their manufacture, transportation, installation, maintenance and the traffic jams that maintenance causes.

By having more efficient appliances, we can drastically reduce our water usage along with the need for sewer. For what little water is still needed, we can collect water on roofs or from the air and store it locally. Composting toilets and graywater gardens can handle the water we don't recycle. Thus we eliminate water pipes, sewer pipes and their manufacture, transportation and installation. With electricity, gas, water and sewer pipes no long under our streets, the ever-present road construction caused by the need to repair underlying infrastructure is dramatically reduced, thus increasing roads traffic-carrying capacity.

Reducing goods transportation means less of a need for traveling salesmen (including business deal makers and ambassadors).

Reducing military transportation

Of course the military can take advantage of many of the above reductions. But there's a special reduction to be made for military. By reducing oil use, and solving scarcity, we reduce war and the need to move an awful lot of people and supplies thousands of miles. However, the military can do a great job in disaster relief because they are experts in logistics. If current trends prevail, there will be plenty of climate-induced disasters. Mitigating our polluting transportation infrastructure will help, but this is a long-term play.

Efficient transportation

We can't eliminate all transportation and still lead a comfortable lifestyle. So for the transportation we still need, high-tech to the rescue!

Personal rapid transit

Cars in the US contain, on average 1.2 people. Moving a 3K pound vehicle with a payload of one 150 pound person (5% of vehicle weight) is wasteful. For urban and dense suburbs, Personal Rapid Transit (PRT) makes sense. It uses guideways over city streets. The best systems have a guideway with a cross section of a little more than a square foot, with 2 passenger pods hanging below. The guideways can form loops or even grids, with stations every ¼ mile to every mile. Pods are waiting at stations for people so that when people arrive they get into a pod and go with no waiting. The pods are automatically routed to the destination station, bypassing the "off line" intervening stations for non-stop traffic. Speed in a city can be 30 to 45 miles an hour, beating single digit bus and car MPH by several times. Since the guideway is much cheaper than subway or light rail lines, and stations are much smaller and cheaper, placing them at frequent intervals in a city means every spot can be within a several minute walk. This is the only transit architecture we know of that is cheap and fast enough to be viable in suburbs (2K residents per square mile and above).

Energy usage is much lower than a car. At several hundred pounds, there's a lot less mass to accelerate. Because the pods travel non-stop, they don't need to re-accelerate at each red light or stop sign. Under computer control, there are no traffic jams or accidents to slow down travel and waste energy. Due to lower frontal area (the two passengers can be in tandem), and a more aerodynamic shape, aerodynamic drag is lower than is practical for cars. Hard wheels on steel rails or even MagLev decreases rolling resistance, further increasing energy efficiency. Electric motors improve efficiency further, is quiet, low maintenance and can be solar powered from panels on the guideway or near by. Hundreds of MPGe are possible with the right engineering, several times more efficient than electric cars.

PRTs can also handle most of the goods transport in cities (see below). The back seat can contain a foldable bike or electric scooter. We can use modified pods for ambulances and trash hauling. Workmen can store tools in lockers at PRT stations so they don't have to lug them the last ½ mile home. PRT stations can have mail boxes. Since residents within a PRT network will likely be using it each day, picking up mail and packages at the nearest PRT station on the way home (you get a text if there's any mail) doesn't cost much time and saves the expense, pollution, noise and danger of delivery trucks.

A well implemented PRT system is cheaper per passenger mile than cars, buses (including bus rapid transit), light rail, and subway. We estimate that a system in a modest sized-city of 150K population or more can be self-supporting on passenger fares that are significantly less than all the above modes with no taxes for capital or operational costs and far fewer externalities. Fry drew up a national plan using PRT that won the Judges choice award in a sustainability contest at MIT's business school in 2012 [Fry 2012]. Two particularly efficient systems are [Skytran 2016] and TransitX [Stanley 2017].

Last mile

PRT stations in a well-covered city will likely be average 1/4 of a mile from your door (3 to 4 minute walk), though some locations could be a mile or more from where you really want to go. A well-designed PRT pod will have enough room for a person and their bike, particularly if its a foldable bike. Scooters, including powered ones, can be even smaller than a bike. Our July 2015 favorite for this "last mile" is [OneWheel 2016]. It is essentially a powered skateboard with one big, central wheel 11.5 inches in diameter, 6 inches fat with a 2 horsepower electric motor in the hub. Top speed is 15MPH, range is 6 miles, charge time is 20 minutes, cost is $1.5K. Because of the big wheel and computer-controlled balancing, you can ride it off-road, so the sidewalk/pavement infrastructure is unnecessary. At 25 pounds it is carriable, but perhaps future versions will allow you to remove half the battery for short-distance commuters. This product is yet another successful Kickstarter project. The last item on their spec sheet: "Awesome: extremely'.

People power

With cars off the streets, it's safer (you won't get run over), healthier (improved air quality), quieter, faster to walk and bike. Bike share enables people to rent bikes for a short period in a city (typically 30 minutes). It is convenient. Since the system maintains the bikes, you don't have to have a place to store it, and you can ride a bike just one way on a trip. Bike share systems can't support themselves off of rider fares. However, by encouraging fewer cars and a healthier lifestyle, in the larger picture, they're a win. [Bike-Sharing 2017] says there were 2.3M public bicycles worldwide in about 1200 cities, at the end of 2016. Vélib, Paris' mature bike sharing system, says that each bicycle is used an average of 6 times a day.

Car sharing

Cars are bad in the city, but to go beyond the economical reaches of a PRT system, there isn't another good alternative. We like the Zipcar model at the edge stations of a PRT system to let people get into the sticks without the expense and space-hogging of owning a car. Since shared cars get used many more times a day than private cars, they save on parking spots, one of the most expensive aspects of cars.

Autonomous automobiles

The Google self-driving car gets a ton of press. It is similar to PRT in that both have electric motors and no driver. PRT is better because it's less accident-prone than a car can ever be due the highly controlled guideway. Because it is not prone to crashes, it needs less weight in crash protection, making it more efficient. It also needn't stop at intersections and does not have traffic slow-downs. This also improves efficiency since a PRT just needs to accelerate once per trip, then goes a constant speed to its destination. Good PRT's also have less rolling resistance due to maglev or hard wheels on steel rails. You could build a 2 seater tandem car, but big car manufacturers don't. Good PRTs do, so they have far less aerodynamic drag, making PRT even more efficient.

Autonomous cars need lots of sensors and computrons for scene recognition. PRTs don't so they are cheaper. The lighter pod makes them cheaper too. A PRT guideway is about the same cost as a lane of freeway but can handle more passengers per mile, and takes up a fraction of a lane's worth of land. The guideway needs less maintenance than a road. PRTs and their guideways are both more reliable and cheaper.

PRT systems need a certain concentration of riders, so they are not cost-effective in rural areas. Here's where cars, including autonomous ones have an advantage. Half of the US population lives in areas dense enough for PRT to make economic sense. The advantage of PRT over cars (of any sort) is much greater than the advantage of autonomous cars over regular cars. Thus PRT is a much better investment of near term R&D dollars, but autonomous cars have received much more investment.

Hyperloop

The 1981 book named *2018* by Gerard O'Niel described vacuum tube "track" for an extremely low aerodynamic train which could travel at thousands of miles an hour using little energy.

In 2013 Elon Musk proposed Hyperloop [Musk 2013]. Hyperloop tubes are not quite a hard vacuum, making them somewhat less dangerous for humans. By directing the small amount of air underneath the vehicle, the Hyperloop vehicle can "ski" on the airflow, with very little "rolling resistance", and as such doesn't need maglev technology. The rough specs for an LA to SF route are: 350 miles, 600MPH, 35 minutes, $6B track cost for $17M per mile. One critic projected a cost of $100B.

There's a lot of clever ideas in Hyperloop. But most traffic is not inter-city, it's within a metro area. Hyperloop is solving a significant, but relatively minor problem. Because of its speed, stations need to be far apart to accommodate their long on and off ramps. Stations are also expensive since they have to carefully handle air pressure.

People don't care about traveling merely station to station, but rather, door to door. 28 passengers per vehicle means that you couldn't hope to put stations near where people were going and maintain "non-stop" travel. This is the same problem that airplane travel has today. Only once we have decent bike and PRT networks, would a system like Hyperloop be worth looking into.

Efficient cars

Modern cars are quite inefficient. We can do much better. In 2002, Volkswagen made a prototype named the 1-Litre. It was a tandem 2 seater, very low aerodynamic drag, weighed just 639 pounds, and had a one cylinder diesel engine that had a measly 8.6 horse power. It got 238MPG. 13 years later, why aren't these cars common? [Arcimoto 2017] is a 3 wheeled, 2 seater, all electric car, with a range of about 100 miles, getting north of 200MPGe. Production is estimated in 2017 for less than $20K per car.

A more radical design is the Lit Motors Motorcycle. It has only two wheels, but it is fully enclosed with an aerodynamic shell and a steering wheel. It has two powerful gyroscopes that keep it up right, even when stopped. The all electric engine and light weight make for a quick bike that promises 400MPGe or so. The range is 200 miles per charge. The Lit motorcycle is so efficient that a foldable solar panel stretched over the top of it while it sits in a parking lot all day will provide enough energy for many commutes. Lit was started in 2010 but as of the end of 2016, production is still a ways off. 3D printing has been used for prototyping cars and other vehicles. As materials get stronger and printers can produce bigger parts, printers are now being used for making much of actual cars.

Airplanes

Big airplanes are noisy and polluting. Their airports take up a lot of room. They're a terrorist magnet, or at least an opportunity for the Security Industrial Complex to rake in more of your money. A 150 MPH, a PRT pod can go the 2.5K miles from NYC to LA in 16.6 hours.

Though that's a lot slower than an airplane's theoretical 500MPH, consider that taking a fast PRT, you wouldn't have to determine in advance to take a trip, spend an hour buying the ticket, travel to an airport, get there an hour early, take off late because they canceled your flight since every seat wasn't filled, wait for 200 people to get on and off the plane, walk long distances in the airport and get from the airport to your actual destination.

Terrafugia

There have been a few personal airplanes developed as of late. Take the "flying car" approach of [Terrafugia 2017] which can travel of road as well as fly. Their "Transition" model get 20 MPG in the air. Terrible gas mileage, but considering it flies at 100 MPH and doesn't need a road, perhaps it will lead to something.

Puffin

If you have to fly, you want to reduce frontal area to an absolute minimum. One person lying down. But you also want vertical takeoff, so your driveway can be an airport. And you want electric motors, so that the noise won't bother the neighbors. That design is NASA's Puffin [Choi 2000].

Twin 30 HP electric motors tilt to allow vertical take off and landing, but its horizontal flight mode is much more efficient than a helicopter. This is still a lot less efficient transportation than PRT or a Lit motorcycle, but if you have to get somewhere quick without roads, this is our favorite.

Efficient transportation of goods

Because PRT is clean and quiet, it can enter buildings unobtrusively. Because it is computer controlled, it doesn't need a passenger to direct it. An assembly line can say "I need more raw materials" and a pod delivers raw materials to the beginning of an assembly line. Also, the assembly line can say "Take away the finished product". If a store has requested more of the good, it can be delivered directly into the back of the store. If not, the good goes to the warehouse, all in the pod. A pod can transport about a cubic meter.

For small stuff, we can use drones. Matternet claims they can deliver a kilogram, 10 km in 15 minutes for 24 cents (including capital and operating costs). A newer version can go 20 km. Good for leapfrogging washed-out African roads. Good for leapfrogging urban congestion. Much less energy per payload pound per mile than a car.

An alternative to transporting packages in the sky, is transporting them on the ground in vehicles that don't need to carry a person. A general article on DeliveryBots is at [Templeton 2016] A particular implementation of this idea is for sidewalk traveling robots by the wrongly named company "Starship Technologies" [Templeton 2015]. The idea is that small autonomous package carrying robots can roll along sidewalks slowly between a supplier and your cell phone.

For big stuff, blimps might be the answer. They can be cheaper and use less energy than air transport. They don't need roads, nor airports, meaning that delivery doesn't need another "mode" to get the product from source to destination [Pasternak 2017].

Driverless package delivery can save people the time to get stuff, save energy because a person doesn't have to be moved, be safer since there's no person in the vehicle and perhaps save on infrastructure.

Conclusion

There are, undoubtedly, a few uses of transportation we didn't cover. If a city or suburb doesn't have many roads, a fraction of the colossal savings can be used for the occasional all-terrain vehicle or helicopter. Modern mountain bikes are pretty capable of rough terrain. We can even reduce the need for roads in rural areas, by using advanced flying machines. Faster travel reduces the time to go places. It also means we can live "further away" without as much a time-penalty, but on cheaper land. Land is the dominate cost of a house, especially once we have 3D printed ones.

But the biggest wins come from just transporting designs on the net, at the speed of light, to a 3D printer that uses locally available raw materials. In any case, traditional automobiles are headed toward the antique museum. Given global warming, it's not a moment too soon.

Chapter 30
Justice

Most people have a deep-seated need to be treated fairly. Without this, they will take revenge against the perceived perpetrator. Perpetrators often think such reprisals are excessive or not warranted at all and take their own counter revenge. Such vicious cycles can continue for years. Especially with gradually escalating conflicts, "who started it" becomes impossible to determine and/or moot.

The concept of a "justice system" was invented to nip such counter-productive interactions from continuing unabated. Over time, these entrenched systems have become primarily monopolistic businesses that are rewarded by creating more conflict. Police in much of the world are funded by the citizens they apprehend. The guilt of the citizens is secondary to the revenue stream they provide. The US program of "cash and carry" [Sallah 2014] proves this kind of pervasive police corruption is not limited to the developing world.

The process

The basic legal process, at least in the USA, is:

1. Apprehend suspects.
2. Determine their innocence or guilt.
3. If they're guilty, imprison and/or fine them.
4. Don't help them return to society when they get out.
5. Since it is difficult for convicts to get jobs and housing, many will resort to crime.
6. Repeat

Little of this process has much to do with Justice. In fact, in promotes more injustice. The best solution is, to the extent possible, to get rid of most of the legal process by eliminating root causes of crime.

Police

The role of the police is to deter and apprehend suspects. By drastically reducing scarcity, we drastically reduce property crimes. By regulating, not prohibiting drugs, another large class of criminal behavior disappears. To the extent that marijuana replaces alcohol, bar room brawls go up in smoke.

With fewer thieves, the need to keep a gun in a house is greatly lessened. For the remaining paranoid, non-lethal weapons (for both police and home-owners) will reduce violence. No one gets shot in a domestic dispute. We can reduce domestic violence even further by providing a place to sleep for a couple of nights for anyone in a crisis.

If we design cars to not be able to exceed the speed limit except for "emergency passing", and to automatically report erratic behavior, most of the need for traffic cops disappears. Hand-eye tests before starting a car can greatly reduce drunk, sleepy or otherwise accident-prone driving. Parking tickets can be mailed out based on a car's GPS, eliminating meter maids. Personal Rapid Transit (See *Transportation* [Ch. 29]) further reduces traffic violations.

With reduced crime, there will be fewer incidents where citizens feel wronged by the police. Understanding the psychology of rage may help threatened police (and others) react with restraint [Shirmer 2015]. We expect that the elimination of war, "class warfare", and much of political polarization, will result in fewer angry political demonstrations, and less need for crowd control.

We'll also have fewer veterans with PTSD. Roughly three times the number of Vietnam vets committed suicide than died in combat [NIMH 2017]. 20% of Iraq war vets have PTSD, a major cause of suicide and domestic violence [Badger 2014].

Most child abductions are by parents that feel they've been treated unfairly in separations or divorces [Robinson 2016]. Divorce is largely about "who gets the money" with the government coming down firmly on the side of ... the lawyers. The primary business strategy of divorce lawyers is to encourage the participants to fight with one another (this is like offering a terrorist a gun so he'll leave you alone). With a decent, lawyer-free justice system, we surmise child abductions and all sorts of aggressive behavior will go down.

White collar crime

Crime might be associated in the public imagination with poverty, but white collar crime is where the big bucks are. Insider trading on Wall Street is pervasive [Grurich 2014].

That kind of money buys a criminal multiple McMansions, yachts, airplanes, ... and politicians and lawyers. Politicians and lawyers are mostly the same group of people, since lawyers become politicians and vice versa. They then promote white-collar crime by resisting laws targeting it, and thwarting enforcement. This makes white-collar crime particularly nasty, undermining public confidence in business and government.

Why do white-collar criminals, usually already well off, do it? We chalk it up to the unbounded-acquisitiveness ethic of Capitalism. Our business and political hierarchies select for sociopathic people who single-mindedly seek wealth and power, in a form of Obsessive-Compusive Disorder (OCD).

Racism

The killing of an unarmed black man in Ferguson, Missouri in August 2014 started a media frenzy of reporting on related incidents in the United States. Many minorities are familiar with systematic racism directed at them by police.

The frequency of such incidents was exacerbated by the philosophy of Broken Windows Policing, which said, *fix the little things and the big things will be fixed too.* New York, under mayor Rudy Giuliani, adopted *stop and frisk,* later found unconstitutional. It quickly morphed into *hassle minority citizens.*

In 2005, Martin O'Malley was the mayor of Baltimore (and in 2016, a candidate for US President). In Baltimore under O'Malley, "in 2005 an incredible 108,000 of the city's 600,000 residents were arrested.".

"Between 2002 and 2004, Chicago police received 10,149 complaints of misconduct, which resulted in only 19 total acts of meaningful discipline". "Between 2009 and the first half of 2014, New Yorkers complained of 1,048 incidents involving choke holds, which had been banned by NYPD for more than a decade". "None of the those offending officers saw significant repercussions" including the policeman that killed Eric Garner, another unarmed black man [Taibbi 2015].

Some muse that "The Criminal Justice System" is so named because the system itself is criminal. We suspect the best fix is reducing scarcity. That will reduce criminal activity and decrease mistrust between inner city residents and their police.

Determining guilt

The primary role of the courts is supposed to be to determine the innocence or guilt of a suspect. Importantly, nobody in a court room actually cares about the truth (except an innocent suspect). The prosecution wants to prove the suspect guilty. The defense wants to prove the suspect innocent. The judge wants to enforce the rules. The jury, if there is one, wants to go home [Bennion 2015]. All this arguing is expensive, and doesn't address the truth. We think the professionals shouldn't compete, but rather should collaborate to find out what actually happened.

Because of the competitive nature of a court room, cases get drawn out. The trick of "plea bargaining" was invented to save prosecutors' effort. The truth is not a commodity that can be bargained. Innocent suspects are browbeaten into copping a plea, because they have no faith in the process. According to one study [Palmer 2013], juries get it wrong 10% to 25% of the time and judges wrongfully convict in 37% of their cases. With those odds, a perfectly innocent defendant may conclude that its better to plead guilty to a crime they didn't do and take a shorter sentence than to proceed to a trial with an uncertain outcome with a potentially longer sentence.

The fact that laymen's judgment is twice as good as legal professionals tells you something important about our "Justice" system. If you had a job where you had to make a critical judgment about a person's future, and got it wrong over a third of the time, don't you think you'd be fired? Its hard to find information on judges getting fired because its so rare.

Cooperative justice

Rather than a court being a battle field where competitors attempt to destroy one another, we propose a different model, which we call *Cooperative Justice*. A suspect is, within 24 hours of being detained, asked if he did what he is suspected of doing. If he says "yes", he's moved into Restorative Justice for sentencing and possibly a Norway style "campus" (see below).

If the suspect says he didn't do it, at first he's believed, and invited to join several professional "truth finders" to help figure out "who dun it". The truth finders explain to the suspect that anyone who is found guilty, the sentencing in Restorative Justice and/or the Norway style prisons are pretty good. This lessens the severity of punishment and lowers the downside of admitting guilt.

The truth finders, assisted by the suspect, examine all the evidence together, including encouraging the suspect to tell them everything he can about the case that will help them all find out who's guilty. The suspect might inadvertently reveal a clue that proves his guilt. He might "give up" and admit guilt, or, together, they might discover another suspect who gets the same respectful treatment. If the proceedings are inconclusive, the suspect is freed, but must be available for further discussion should new evidence or suspects turn up.

Sentencing

Sentencing is designed to punish the convicted. This gives the victim some sense of "justice", but does little to prevent a criminal from re-offending. Recidivism is the rate that a criminal is convicted of another crime after they leave prison. In the US, it's 76% over 5 years [Dickinson 2014].

Restorative Justice is a fairly new process whose goal is to prevent recidivism while giving the victim a more concrete compensation. The victim, the criminal and their friends and family, are brought together to discuss the appropriate "restitution" that the criminal should pay the victim. It could be money, time working on a project or some other thing the victim believes is valuable. The criminal must apologize and give the victim their due. Lots could go wrong with such a process, but experience shows that more goes right than in the conventional process [Sherman 2007]. Restorative Justice is mis-named because it only deals with sentencing, not determining guilt. Still, we believe it can be an important piece of justice.

The prison industrial complex

The Prison Industrial Complex (PIC) is an industry that benefits from convictions, the more the merrier. They lobby to keep drugs illegal, because 65% of US inmates are drug abusers. Combine those with drugs being involved in how they got into prison in the first place, and we're up to 85% of the prison population [Sack 2014]. The PIC benefits from repeat customers so they have an incentive to increase recidivism. "As it is, our prison system does little more than teach addicts how to be better addicts." [Sack 2014]. Prisons are "universities" for learning new tricks of the trade. Parole officers are supposed to counsel ex-cons in how to stay clean. They're admittedly understaffed, but they're also grossly ineffective.

Norway has an entirely different model. The purpose of prison is to prepare a criminal for integration into society when they get out. Since Norway has no death penalty and a sentencing limit of 21 years, most criminals do get out. The basic strategy is to have the prison be as normal and supportive as it can be. Prisoner's "cells" look like private college dorm rooms. They have a door that the prisoner chooses to leave open or closed. There are kitchens where prisoners can cook for themselves or others. They take classes in useful skills. They have jobs where they can earn money, some of which can be spent on the inside on clothes, special foods etc. and some is saved to help the prisoner when they get out. They are not released until they have a job and a place to live.

From [Benko 2015] we learn: US prisons cost $31k per year per inmate. This Norwegian prison costs $93K. But proponents claim it's worth it. Recidivism in the US is 60% over 2 years but in whereas in Norway its only 20% over 2 years. (Due to other differences, it's prudent not to read too much into such statistics.) Instead of a repeat offender, or an unemployed or homeless person, society gets the benefit of the skills of an employed citizen in a stable situation.

Stop competing for justice

The major theme of this book is that cooperation is better than competition, especially if we solve scarcity. As in most of our institutions, competition is the dominant process in the US legal system. It imposes *artificial scarcity* on suspects for months or sometimes years with a policy that is effectively "Justice delayed is justice denied". Our courtrooms can't hope to find the truth with expensive, adversarial lawyers, plea bargaining, inconsistent application of (lobbyist-written) law, intimidation, etc.

New processes for justice can benefit everyone (but the current legal professionals). Cooperative Justice is a plausible new way to better understand the truth behind crimes. Restorative Justice helps victims recover and offenders appreciate the consequences of their actions. Norway-style prison campuses treat convicts with respect, and train their "students" in how to make it without crime. "Parole" should help convicts find jobs, housing and a more responsible lifestyle rather than merely restricting their freedom.

The battlefield of adversarial justice makes the process of the US legal system more like war than science. Because

> *"Truth is the first casualty of war"*
> *– Sen. Hiram Johnson, 1817*

truth is also the first casualty of our legal system. And prisons are just like military field hospitals, in that as soon as as the residents leave them, they're sent into activities likely to land them back in the same place, in short order. Justice isn't supposed to be a battle—it's supposed to end battles.

Chapter 31
Guns

Few topics are as polarizing as guns in American politics. In 2010, there were 19,392 firearm-related suicides, and 11,078 firearm-related homicides in the U.S., totaling roughly the same as traffic accidents. Then you must add in the gun injuries that are several times more. There is roughly a gun per person in the USA.

Background checks, often proposed, are an ineffective solution. An awfully high percentage of the population has pretty good control of themselves nearly all the time. It's that "nearly" part where the fatalities mount up. Too often, also, the buyer is not the user.

Hunting

Even if you endorse hunting for food or for sport, how can we make sure that little Johnny doesn't take that sporting gun to school? Leaving it home doesn't work, so let's keep it at a sporting club. An owner could "check it out" on their way to the duck pond and bring it back before going home. With GPS tagging of the car, the shooter, and/or the gun, we can have a pretty quick notion of the gun's likelihood for being used for something other than sport.

Home defense

Let's solve the "home defense" use case with non-lethal weapons. Statistically, it's far more likely that a family member will be shot than an intruder. From [Kellerman 1998], "For every time a gun in the home was used in a self-defense or legally justifiable shooting, there were four unintentional shootings, seven criminal assaults or homicides, and 11 attempted or completed suicides." When Johnny accidentally shoots his sister with a mace gun, it's a trip to the emergency room, not the funeral parlor. We've got Tasers, sticky foam, various chemicals, bean-bags, stun grenades, slow bullets [Atherton 2015], and more research will yield more options.

Some gun owners might think they need a gun for defense against animals. One of the above non-lethal technologies might be appropriate. Perhaps not as convenient as a gun, but way more convenient than a funeral.

Some might also imagine that guns would have a use in defense in public gatherings. Imagine you were at the rally where US Rep. Gabby Giffords was shot. Imagine that lots of people there had concealed handguns to "prevent such things from happening". You hear a shot and everybody pulls out their gun to shoot the bad guy. Hmm, which of those people now brandishing a gun was the original shooter?

Car defense

A driver might be legitimately concerned with car-jacking. Our transportation solutions minimize this. One alternative for cars is a hidden "kill switch". This might make the car-jackers mad enough to harm you. How about if the kill switch lets the car run for a mile, while transmitting the car's GPS to the local police?

There's another use case for a gun in a car: shooting the driver that just cut you off. Given the human psychology of road-rage, uncontrollable anger while driving is all too common. But you don't have that uncontrollable anger when you get in the car. Use that wisdom to decide not to take the gun with you.

Police misuse

Non-lethal weapons are also appropriate for police. Police can make mistakes like anyone else. Reducing lethal mistakes will build public confidence in the police.

Defense against the government: militia

A use case cited by some is a "milita" against the government. OK, this "worked" in 1776, but note that Canada didn't need a war with Britain, and it's done pretty well. The ratio of firepower of government to citizen has gone up considerably in the last couple hundred years so "militia" isn't a realistic strategy for the tyranny of government. We propose solutions to abuse of power in the *Government* chapters [Part 4].

Anger

Consider: "Nearly 9 percent of the adult population in the US have impulsive, anger issues, break or smash things and get into fights - and have access to at least one firearm" [RT 2015]. But it is hard to identify those 9 percent. We need better mental health services, but the surest way to keep guns out of the hands of impulsive people is to keep them out of hands. Period.

Drug dealers

Drug dealers have a tendency to own and use guns. By making drugs legal but regulated, we drastically cut one of the deadliest uses of guns. With a better Justice system and prisons designed for the inmates to constructively re-enter society, rather than just punish them, we reduce gun-ownership incentives even further [Sterbenz 2014].

Freedom

We agree that restricting ownership is a removal of one sort of individual freedom. But giving everyone easy access to guns reduces everyone's freedom. When you've been shot, your freedom drops to zero, rendering this argument self-defeating.

Places with fewer guns have fewer gun deaths. Areas with strong anti-gun laws that nonetheless have a lot of gun violence (such as DC and Chicago) are adjacent to areas where access to guns is easier [Gordts 2015].

The 2nd Amendment to the US Constitution is the rallying cry for the National Rifle Association, funded by gun manufacturers who profit from death and fear. They treat this amendment as sacrosanct, but why can't it be changed, as we repealed the alcohol Prohibition amendment? Isn't changeability the very idea of being able to amend the Constitution? Ultimately, though, we need to form a more civil society. That's the goal of this book.

We're at an impasse with guns in America. Perhaps reasoning framed by a comedian will work where logical persuasion has failed? [Jeffries 2015].

Chapter 32
War crimes

The two words in the title of this chapter are redundant. Every war is a crime. Everything that is bad about crime takes place in war: People die. They get assaulted and raped. People are forced to live under horrible conditions. Houses are destroyed. Possessions are stolen. War is portrayed by the governments and armies who conduct it as a necessary evil. In fact, it's just evil. Soldiers aren't heroes, so much as they are victims of the war machine.

Militarists argue for the necessity of war with statements like "war has existed throughout human history" and "the world is a dangerous place". We call bullshit. Humanity has put an end to other barbaric practices of ancient societies such as human sacrifice, cannibalism, and (almost) slavery. There are dangers in the world, but surely dangers can be reduced by methods other than war, cooperation foremost amongst them.

Arguments for the supposed necessity of war have been eloquently refuted, in books like David Swanson's *War is a Lie* [Swanson 2010]. We also recommend reading General Smedley Butler's 1935 essay, *War is a Racket* [Butler 1935], written by the head of the Marine Corps in World War I. Guys like him should know.

The war frame and the crime frame

It matters whether we label violence perpetrated by governments, armies, and political groups as war, or as crime. Linguist George Lakoff [Lakoff 2014] talks about *framing*: explaining a given situation by making an analogy to a simplified, prototypical story that has particular roles and events.

Which frame you choose determines what underlying assumptions you will make. Politicians and business people often promote misleading frames in order to manipulate the way people view situations.

War has a frame: it has opposing armies, enemies, battles, victory or defeat. Crime also has a frame: it has criminals, victims, suspects, police, trials, verdicts, and sentencing. There is a subframe of crime called organized crime, where crime is perpetrated by a long term, large-scale group that has significant resources and the ability to plan and execute criminal activity.

One of the big differences between the war frame and the crime frame is what you can do about it. In the crime frame, you conduct investigations that identify individuals, collect evidence, follow the money to determine how criminal activities are financed, etc. You can't commit violence against a suspected perpetrator unless you have evidence. You have to respect the rights of the suspect.

In the war frame, you shoot. You try to destroy the enemy. No rational thought or proof about whether the suspect is guilty of a crime is necessary. The actions of "our side", even if violence or destruction is involved, are usually not questioned or always assumed to be justified in self-defense. Mistakes that come to light are "collateral damage", or "friendly fire", to be excused.

Another big perceived difference between the frames is that war is an emergency, and crime is generally not. The key thing about an emergency is that it is a very serious threat, and there is a scarcity of time for the response to it.

As the *Scarcity* book [Mullinaithan 2013] tells us, scarcity brings both focus (perhaps positive) and blindness to consequences. Scarcity also disrupts rational thought. In order for the military-industrial complex to get the public to ignore the negative consequences of war, they have to portray every military situation as an emergency.

To sum up, "war" and "crime" are just different ways of talking about the very same thing: violence against innocent people. The differences are whether they are sanctioned by government, and perhaps that war has a larger scale than crime.

But it's worth paying attention to which frame is used to talk about it—the crime frame gives us a path towards constructively dealing with it, whereas the war frame precludes any real solution.

Why did the assassination of Archduke Ferdinand start World War I? Sinking the battleship Maine start the Spanish-American War? Why did 9/11 spark the wars in Afghanistan and Iraq? Because these crimes were framed as acts of war, leading to a runaway feedback loop of escalation.

Does the war frame make sense for combatting crime?

Long-term solutions, like improving education and the economy in the third world, are rejected out of hand because they aren't perceived as having the potential of an immediate fix, whereas (despite the evidence), war is.

Never mind that the 12 years and $4 trillion spent on the Iraq war were certainly long-term and large-scale investments, and they failed. In order to avenge the deaths of about 3000 Americans in the September 11 attacks, the Iraq and Afghanistan wars caused 5000 American deaths and hundreds of thousands of deaths of Iraqis and others. While things like tightened airport security may have reduced our vulnerability to further attacks, the military actions may have increased the total number of jihadis willing to conduct such attacks. Was that rational?

Thinking about today's situations of endemic violence in the Middle East, perpetrated by groups such as Al Qaeda and the so-called Islamic State, should we view this as war, or as criminal activity?

Think of Al Qaeda or ISIS as being like the Mafia. They're organized crime. Neither Al Queda nor ISIS is a government (though they claim to be), nor an army, really. Actually, far more people work for the Mafia worldwide, than ISIS. The Mafia's annual budget is vastly greater than money spent by radical Islamic groups. Probably more people die each year as a result of Mafia activity than in Middle East wars.

Yet nobody suggests bombing Sicily or New Jersey as a way of combating the Mafia. Instead, fighting the Mafia is done with methods meant to deal with high crime. In the UK at least, the organization charged with fighting terrorism is part of Scotland Yard, the crime-fighting organization, not part of the military.

Any opponent of war encounters almost immediately the objection, "What about World War 2?" Was it justified to call that a war and not a crime? Was it, as Studs Terkel called it, *The Good War* [Terkel 1984]?

Certainly, it came the closest, of any historical event, to fulfilling the frame for war. It was instigated by the national governments of Germany and Japan, each with an organized army. It truly was an emergency, as Hitler's armies were marching through Europe, and Japan attacked the US at Pearl Harbor. And never before had such violence against innocent people been unleashed at such a large scale. It's hard to have any sympathy at all for Nazi Germany and Imperial Japan. So if anything qualifies as a just war, that one does. We're willing to concede this one example.

But we still can't help but feel that, had the long-term solutions we are advocating—relieving material scarcity, teaching people to be more cooperative, improving mental health, developing more rational decision-making processes for government and other organizations—been in place, they might well have stood a chance of preventing the situation from deteriorating until war was inevitable.

Suppose a mental health professional had been able to provide therapy for Adolph Hitler (who was beaten by his father as a child), when he acted out as a troubled teenager? Suppose the economic troubles and runaway inflation of 1930's Germany (themselves a consequence of World War 1) hadn't happened? Would people have elected Hitler? Suppose Germans had been better educated about science, so they would have recognized Hitler's ranting about eugenics in *Mein Kampf* as insanity? This is all in the realm of *shoulda, woulda, coulda,* so it's impossible to know. Chapter Four of David Swanson's book, *War is a Lie* [Swanson 2010], treats World War 2 and debunks in detail conventional arguments for its inevitability and its "goodness".

After World War 2, the US government at least had the foresight to realize that poverty from World War 1 was a major cause of World War 2. If it left Europe in ruins, it risked yet another war. So it launched the Marshall Plan, helping Europe to get back on its feet. Where was the Marshall Plan for the Middle East after the first Iraq war? So guess what happened…

After seven decades, Germany is at "war" again—this time, beating the US—in installation of renewable energy sources. And thanks to Germany's influence, the US finally has decent beer. Let's move national pride from conquest to culture. Battle of the Bands, not Battle of the Bulge.

We should stress that we have no personal animosity towards the rank-and-file men and women of the armed forces. Our scorn is reserved for war-mongering politicians, the military-industrial complex, and those elements of the military that do promote or exacerbate war, or commit gratuitous violence. We realize that soldiers often join out of an admirable sense of selflessness, discipline, and willingness to defend others. If they join out of perceived economic necessity, that's sad; they are victims of a dysfunctional economic system.

We would counsel them that the military isn't the best way to fulfill their desire to serve—instead, join the Peace Corps, Médicins sans Frontières, or any one of a number of groups where you can help others, travel and have adventures, learn discipline and camaraderie, and make friends for America rather than enemies. Feel like you have to serve by having a gun and threatening violence to bad guys? That's great, too—every police department in the country is hiring, every year (though we'd prefer they purchase only non-lethal weapons).

In the US today, there is increasing strife between local police departments and low-income communities, such as in Ferguson, Missouri, New York City and Baltimore in 2014-5. One of the reasons for this has been that increasingly, fighting crime in urban areas has been reframed as a "war on crime".

Military hardware has been sold to local police departments, an economic boon for the military-industrial complex. Taxpayers pay twice for this, once when the military buys it, and again when it is sold to police departments (and a third time, when they have to deal with the consequences of weapons use). Peaceful street protests are met with tanks and military weapons. Police departments are staffed with returning Iraq and Afghanistan war veterans, many with PTSD. Rights of criminal suspects are ignored. No wonder many residents of low income communities feel like they are being invaded by an occupying army.

The acts that provoke wars, like the September 11 attacks, are evil. But they're crimes, not acts of war. The invention of the linguistic frame of war is perhaps the greatest crime ever committed.

Chapter 33
War

"When you're tempted to fight fire with fire, remember:
The pros use water." – Kristin Hall

Of all the problems besetting mankind, war is likely the worst, perhaps simply because we "do it to ourselves".

We dive deeper into the fallacy of "wars between nations" through a series of fictitious interviews that emphasize the diversity and individuality of motivations. We're caricaturing here, so please forgive us our oversimplifications. We trying to show that, while "nobody wants war", seemingly reasonable motivations can add up to global, historic tragedy.

The citizen

Mr. Citizen lives at 123 Main St, Anytown, USA.

Interviewer: Do you consider Ahmed Ahmed your mortal enemy?

Citizen: Who?

Interviewer: Ahmed Ahmed, a citizen of Fooistan.

Citizen: Never heard of him

Interviewer: Then why did you pay to have him killed?

Citizen: Did not.

Interviewer: I have here your last year's tax return. You willingly forked over $16,678. (oddly, exactly the average income tax plus payroll tax 2013 for a single worker) [Lundeen 2014]. One third of that [Claremont 2016] went to veterans and the military who blew up Ahmed Ahmed's house on February 12th.

Citizen: It wasn't willing. If I didn't pay my taxes, the government would throw me in jail. Then, without a job, the government would confiscate my house for back taxes, take the proceeds and blow up Ahmed Ahmed anyway. I had no choice.

Interviewer: Well why didn't you vote for a peace candidate then?

Citizen: I voted for the most peaceful candidate running, who escalated the forever war. The other guy would have been even worse.

Interviewer: Well why didn't you support a MORE peaceful candidate in the primaries?

Citizen: I did, but he lost because he didn't have the big campaign donations allowed by Citizen's United. Only the big companies can give that much money. They get the money from profits of selling weapons, and use part of those profits to fund politicians to ensure we have war, i.e. a market for weapons.

Interviewer: Well in the end you contributed to Ahmed Ahmed's death.

Citizen: Did I have a choice?

The President

Interviewer: Why did you cause the war?

President: I didn't. My predecessor did.

Interviewer: We asked your predecessor and he said the 9/11 terrorists started it. So we asked the relatives of those terrorists and they said they had no choice since America was destroying their culture. Plus America gave weapons to the other tribe to beat them up. We kept following this chain all the way to

before recorded history and determined that nobody claims
to cause war. But it is still an entirely human process.

President: I had to bomb Ahmed Ahmed. The other party was using
the politics of fear to whip up the voters, who would have
voted for the other guy, who's even more of a war monger
than I am.

Interviewer: Well, that was maybe true for your first election and
re-election, but now you've got no more elections to
appease fearful voters of.

President: Yeah but I've still got the Joint Chiefs breathing down my
neck. I had to stop them from nuking Iran last month.
Murdering Ahmed Ahmed was a small price to pay. We
had to do *something* after the 9/11 attacks, by God.

Interviewer: You had to do something. Well, what was it that you *had* to
do?

President: Well, we couldn't just let it go.

Interviewer: You didn't answer my question. You said, "we *had* to do
something". What, exactly, did you feel like you were forced
to do?

President: Actually, we weren't forced into any *particular* action. But
we felt like we had to take some sort of military action
against the perpetrators of these attacks.

Interviewer: I call that *doism* (pronounced "do ism"). It's when you feel
like you have to do something, but you don't have any
clear idea of what to do. So you pick some action, without
really understanding what it is supposed to accomplish or
whether it stands a chance of working. That usually ends
badly. What were you trying to accomplish?

President: If we just let the 9/11 attacks go by without any military
response, we'd be vulnerable to future attacks.

Interviewer: So we didn't "let the attacks go by", as you say. And we had
a war. Are we vulnerable to future attacks now?

President: We haven't had major terrorist attacks since 9/11. Unless you count the Boston Marathon, San Bernardino, the Orlando gay club, and a few others. And of course we can never be certain some attack isn't out there being planned. We certainly still have vulnerability to future attacks.

Interviewer: So, in terms of vulnerability, I guess it didn't work.

President: But we destroyed the guys who attacked us on 9/11.

Interviewer: The 19 guys who did it died in the attack, sure. Their deaths were by their own design. Hundreds of thousands of Iraqi and Afghan people were killed in the resulting war, almost all of whom were in no way involved in that attack and would not have planned other attacks. Maybe you did get a few who might have attacked, but do you have any real reason to believe that war made us safer, or that it was worth the loss of human life of that magnitude?

President: I suppose I can't convince you of that. As commander in chief, starting a war meant that nobody could accuse me of not taking action. They usually don't ask me the question you just did.

The Secretary of State

Interviewer: Tell us about your career.

Sec: Well I wanted to be President and my family had tons of money so I went to Ivy U. Then I knew that presidents needed to have a military record so I volunteered for the army. But of course, having a record of real combat is even better on a presidential resume so I requested to go to Vietnam. But I became disillusioned with that war and when I got back to America, I protested it.

Interviewer: Many observers saw the war in Fooistan as just another Vietnam, so why, as Congressman, did you vote to fund Fooistan and the Iraq wars?

Sec: Good question.

Interviewer: And, as Secretary of State, you backed the "forever war" policy. Isn't that contradictory to your protest of the Vietnam War?

Sec: No comment.

Interviewer: Why did the US blow up Ahmed Ahmed's house?

Sec: We had to do something after the 9/11 attacks, by God.

Interviewer: So you had to do something, but you didn't know exactly what or what it was supposed to accomplish. I call that… oh, never mind. What was it you felt forced to do?

Sec: We had to teach those guys a lesson.

Interviewer: So what was the lesson you were trying to teach?

Sec: That if you mess with the United States, you die.

Interviewer: But the attackers wanted to mess with the US, and they also wanted to die in the attempt. So that doesn't seem like it would be helpful in preventing future attacks like 9/11. What lesson did you think they could learn from the Iraq and Afghan wars?

Suppose somebody thought you were doing something wrong, even though *you* thought you didn't do anything wrong. Then they attacked you. Would that cause you to change the behavior your attackers didn't like?

Sec: Of course not! If I wasn't doing anything wrong and got attacked, I'd fight back to defend my honor!

Interviewer: I see. Why do you think the 9/11 attackers did it?

Sec: Because they hated our lifestyle, our religion, our freedom and prosperity.

Interviewer: So they attacked us. They thought they were teaching us a lesson. Did we learn that lesson? Did we change our lifestyle, our religion, our prosperity?

Sec: Of course not! It just strengthened our resolve to defend our values and attack back!

Interviewer: So, when we attacked back, do you think our adversaries learned the lesson that they shouldn't attack us? Or might it have strengthened their resolve and led them to plan future attacks?

Sec: That's up to them. But we had to do something.

The General

Interviewer: Why do you support the war?

General: We have to protect American interests.

Interviewer: Such as?

General: Well last week our boys were driving down a road in Fooistan when an IED blew two guys into the hospital.

Interviewer: Why were they driving down the road?

General: We were going to blow up the house of terrorist Ahmed Ahmed.

Interviewer: Because?

General: He is a suspected IED maker that's hurt our troops all over the region.

Interviewer: But if our troops weren't in the region, they wouldn't get hurt by any IEDs and you wouldn't have to protect them.

General: What's your point?

Interviewer: Sending troops to protect troops is a self-fulfilling prophecy.

General: But we're preventing the next 9/11.

Interviewer: Many people think you're ensuring it by making more people mad at the US. Occupying troops are bound to cause resentment, and they do.

General: This interview is over.

The Congressman

Interviewer: Why did you add more money onto the military budget than even the Pentagon requested?

Congressman: Because Military Industrial Corp. is in my district. They paid for my campaign, which is how I got here.

Interviewer: But that's corruption.

Congressman: That's America. Look, if you want to get anything done around here, you've got to play ball.

Interviewer: Well what have you got done?

Congressman: I got 123 jobs in my district making bombs to blow up people.

Interviewer: But then those bombs actually blew up people. Thousands, in fact.

Congressman: What's your point?

Interviewer: If you have to do bad stuff in order to get elected, then it's better that you not get elected.

Congressman: Well if I didn't get elected, the country would have drugs, gay marriage and abortions and gone to hell in a handbasket.

Interviewer: But the country your 123 workers helped bomb, *did*, in fact, go to hell in a handbasket.

Congressman: Look, I was elected to protect the interests of my district. If that happens to conflict with the interests of Fooistan, well, they're not in my district.

The Recruit

Interview: Why did you join the Army?

Recruit: I couldn't get any other job.

Interviewer: But how about going to college and train in some needed skill?

Recruit: Like what? My best friend went through all 4 years, graduated, and he's still unemployed. I racked up 2 years worth of student debt and didn't want to end up like him.

Interviewer: So you thought getting paid to kill people was your way out?

Recruit: Beats McDonalds, but even they haven't been hiring in my city.

The Terrorist

Mr. Ahmed Ahmed, formerly of 123 Akbar Lane, Warizbad, Fooistan.

Interviewer: Are you Ahmed Ahmed of Warizbad?

Terrorist: Who's asking?

Interviewer: I'm Cory Correspondent from the New York Places.

Terrorist: Ah, you must know my cousin Ahmed Ahmed-Ahmed. He lives in Brooklyn.

Interviewer: Well Brooklyn's a big place...

Terrorist: So what brings you to heaven?

Interviewer: Our readers are incredibly curious about what motivates terrorists.

Terrorist: So why are you asking me?

Interviewer: Well you did make IEDs, did you not?

Terrorist: And Colt makes M16s. Do you call them terrorists?

Interviewer: Well Colt supplies the military with weapons to defend the USA's national interests.

Terrorist: And I supplied the People's Free Army with weapons to defend Fooistan's national interest.

Interviewer: So your contention is that you're not a terrorist?

Terrorist: Less so than your Military Industrial Complex. After all, the People's Free Army didn't invade Brooklyn and start blowing up houses.

Interviewer: The World Trade Center got blown up a few years ago.

Terrorist: By guys I don't know and didn't agree with. My cousin almost got hurt delivering falafels near Wall Street that morning.

Interviewer: Then why did you make IED's?

Terrorist: I'm a baker, just like my cousin. But what would you do if tanks from some foreign country started rolling down your street?

Interviewer: I'd try to scoop the Post on the story. But I see your point. By the way, is it true about the 17 virgins?

Terrorist: Heck no! I only got 14 and one of them was my aunt Fatima.

The terrorist's brother, Demha Ahmed

Interviewer: I see a bunch of bomb parts on your table. Are you following in your brother's footsteps?

Demha: Somebody's got to avenge his death.

Interviewer: Actually no, and the world would be a lot better if we stopped this *avenger* behavior. Note to Hollywood: thanks for not helping.

Demha: You want me to do nothing when your government stomped on Fooistan and my family?

Interviewer: Well, *doism* is a prevalent cognitive bias. And I admit the US president was guilty of it after 9/11, blowing up people that didn't cause the problem. How about if you take over your brother's baking business instead?

Demha: How about you Americans go back to Ferguson and blow up your own people?

Interviewer: Ouch! We already tried that, and judging by the unrest in Baltimore and elsewhere, it didn't work out too well.

Demha: And it's not working out too well here in Fooistan either. The Americans are still here, and every village still lives in fear that it'll get attacked. By the Americans, by ISIS, Syrian forces, or someone else. It doesn't matter who.

To tell you the truth, I wasn't so excited about joining Al Queda in the first place. I'm not even that religious. I was happy in my old job baking pita bread.

But I'm the eldest son of Warizbad's most prominent family. Now, nobody can accuse me of not taking action to defend Warizbad's honor. They usually don't ask me the question you just did.

Interviewer: Turns out, one of the people your bombs blew up was Pfc. Ahmed Ahmed-Ahmed, a falafel vendor from Brooklyn who happened to be your cousin. He joined the Army after the 2008 financial crisis, when a Wall Street bank foreclosed the mortgage on his falafel restaurant in Bay Ridge, and he couldn't get a job. The Army was really happy to have him, because he was one of the few who spoke Fooish.

His platoon mates vowed that they would avenge his death. After all, they *have* to do something.

The Vet

Interviewer: What's the worst thing about war?

Vet: There's so many bad things it's a tough call. But for me, it's coming back and realizing I killed somebody's husband, somebody's son.

He wasn't some faceless terrorist. He was Ahmed Ahmed, a baker for a small shop who became radicalized by the American presence. I wasn't part of any solution. I was part of the problem. I hate myself for that.

Interviewer: Would you say you have PTSD?

Vet: You name it, I've got it. Shrapnel in my leg, guilt in my heart, broken dreams in my head.

Interviewer: Do you have any advice for recruits?

Vet: Do yourself, your country, and the world a favor. Don't join the military. I know it's tough being poor. I know shooting machine guns can feel like a video game. Getting out of your ghetto and going half way around the world is exotic. But all that is small potatoes compared to its cost. I will never recover from war and neither will my buddies and the guys on the other side. If you want to help your country and travel, volunteer for the Peace Corps.

Chapter 34
Deploying Innovation

Yankee ingenuity

America has a wealth of valuable assets. It has, relatively, a lot of land per person (7 acres). It is rich in natural resources, and has numerous ports on 3 coasts, extensive internal pipeline, transportation, and communication networks. The financial sector is strong and American citizens are themselves a huge market. The US maintains a university system that international students, professors and researchers flock to.

Perhaps most importantly, we have a tremendous wealth of innovation. The US is not alone here. Even very poor nations innovate, often more appropriately for their own problems than an outsider could. Nonetheless, many of the most significant innovations of the last century came from the USA.

Why are we stuck?

Despite all those advantages, America appears, to many observers, to be stuck. The American dream seems to have unraveled.

Our GDP may be up in the last couple of years, but median income is not. Job prospects for college grads are slim, and younger non-grads' fate is worse. The percentage of a middle-class person's income that must be spent on the necessities of housing, food, and health care is going up, when technological progress should result in that percentage decreasing.

In 2010, US vice-presidential candidate Sara Palin, in commenting on Obama's 2008 core campaign promises, asked the rhetorical question "'How's that hopey, changey stuff working out?" [Gonyea 2010]. At the time it sounded to many on the Left as just plain stupid, but history has shown its sentiment to be an accurate prediction. We need more than just hope and a desire for change.

Two steps forward, but how many steps back?

The march of science and technology progress goes on, and even accelerates. But does this translate into improvements for most people in society?

Our climate is less hospitable to life, as is our politics of fear and war. Only 27% percent of Americans think the country is heading in the right direction [Rasmussen 2017]. Our national energy policy can be described either as "non-existent", or as "Big Oil". Take your pick. Gridlock is the politest way to describe Congress. The Supreme Court that has declared "corporations are people". Our police departments routinely use the "cash and carry" money confiscation schemes to fund themselves and unnecessarily lethal force. A kid shooting family members in their home is so common that it is no longer a shocking news story. Our Prison Industrial Complex is second to none. The trillions of dollars the US has spent this millennium on "the forever wars" have arguably made the world, less safe. What gives?

Conventional mechanisms for innovation

The conventional wisdom is that our society has built-in mechanisms for fixing itself. If you're willing to "go through the proper channels", we are supposed to have self-correcting procedures that allow innovation to take place. Problem is, those mechanisms have gotten rusty.

- *Elections* are the means which democracy is supposed to fix itself. We see a decreasing percentage of eligible voters voting (36% in 2014 elections). This means it takes 1 more vote than 18% to elect our leaders. Those that do vote tend to do so against their own self-interest. More on problems with voting in the chapter *The Trouble with Voting* [Ch. 23].

- *Citizen protest.* The latest significant ones as of this writing were Black Lives Matter of 2014-5, and Occupy Wall Street, in 2011, spreading to 951 US cities and numerous countries. Obama even said "We are on their side" [Halper 2011]. Then Obama signed into law an anti-protest bill that makes it easier to criminalize protests. [Lithwick 2012].

- *Liberal democracies* allow citizen protest to take place as a harmless "escape valve". But government and corporate officials don't take citizen protest alone as signaling a necessity for change, unless backed up by votes, and by money.

- *Structural change* of government takes a constitutional amendment. For practical purposes, we cannot amend our constitution. It is dead. With a dynamic technical, social, economic and physical environment, a dead constitution will lead to a dead country.

- *Startups* are supposed to be the cutting edge of Capitalism. Since 90% of startups fail, we have the efforts of 90% of our innovation workforce pretty much wasted. Great for the venture capitalists, who make money off of the proceeds of the successful 10%, but for the individuals who don't cash out, tough luck.

And tough luck for society in general, which doesn't get to learn very much from the failures. In science, it's OK for 90% of projects to ultimately fail, because the scientific community learns along the way, by publication and education activities.

But most startups are secretive, so nobody outside the startup learns anything. The participants in a failed startup are usually too burned out at the end to have much more energy for further developing their ideas or for teaching others. Some go on to use what they learned in other startups, but most tend to desert innovation, looking for more stable jobs. The sale of intellectual property to buyers of failed startups is a very poor mechanism for recycling their potential for innovation.

We (the authors of this book, the country, and the world) have many innovative solutions that show promise in tackling our big problems. The roadblock is *deploying* innovation. It's time for something new.

Heuristics for self-improvement

Let's characterize these mechanisms using the language of computer science. Artificial intelligence studies which are procedures that are "rules of thumb" *heuristics*, used for perhaps only partial solutions to problems. They don't always work. If you don't have a total solution to the problem, they're better than nothing. Sometimes they can be stepping-stones to more effective solutions.

When you design government or business structures, you have to design heuristics for making decisions. You can't make all decisions in advance, so the best you can do is design a decision-making process that makes it more likely that good decisions will result. (Our candidate in the chapter *Reasonocracy* [Ch. 25].)

Search algorithms are a kind of heuristic. They are methods for finding solutions to problems in a search space, a universe of possibilities. AI also studies machine learning, another kind of heuristic, by which a program can become better over time, without needing to be explicitly redesigned.

In the language of AI, then, elections, protests, laws, amendments, and startups are all heuristics (processes) for improving society. They don't always work, but they increase the probability that the system can improve itself. Especially when compared to a more rigid system without mechanisms for change, like dictatorships or feudalism.

Generate & test and hill climbing

As heuristics, they mostly fall into two well-known categories: *Generate & Test*; and *Hill Climbing*. In Generate & Test, you have two processes. The first process, *Generate*, spits out possible solutions. The second process, *Test*, tries to determine whether each possibility is good or not.

In evolution, mutation is the Generate process and natural selection is the Test process. In the innovation ecosystem, though a lot of planning goes into a startup, from the larger economic perspective, we could model entrepreneurs as the Generate process, and venture capitalists as the Test process. VCs weed out the implausible startups and fund the promising ones. After funding, incremental steps of success in the marketplace form another kind of Test process. In US Democracy, running for office and proposing laws are Generate processes. Voting is the Test process.

Generate & Test, alone, is known to be pretty weak as a heuristic. If something's wrong with the Generate process, it'll just keep spitting out proposals that fail the Test. If something's wrong with Test, it won't be able to tell the difference between good and bad proposals. But even if we *do* find a decent solution, G&T alone can't determine *why* something succeeded or failed. Modern machine learning has far more sophisticated procedures for generating and testing, that use feedback to improve both processes.

Hill Climbing is a way to make incremental change. Since big change is usually risky, you try a few ways of making small changes and see which one improves the situation. Then you look for another small change. Modern business and politics also use Hill Climbing. Usually, electing one more candidate, passing one more law, introducing one new product, or starting one more company doesn't change the whole system very much (there are notable exceptions).

Entrepreneurs and political radicals may propose new and risky stuff. But the job of established businesses and politicians is actually to avoid and minimize risk, as much as possible. What they really like is Hill Climbing—minor tweaks to already-successful models, and models that can succeed a little bit at a time, then use that success to grow larger. That's why we get so many lookalike startups and lookalike politicians.

Unfortunately, Hill Climbing is also a pretty weak heuristic. It has the nasty problem that you can get stuck in a *local maximum*. Hill climbing means taking incremental steps up the hill (continual, but minor improvements), and never down, until you get to the top. But once you reach the top, you can discover that there were other, higher mountains around you that you can never scale, because it requires first taking steps down before you step up again.

Hill Climbing also has the problem that it has to take just one small step at a time. It is mostly good for incremental change, but not so good for innovation or more fundamental change. AI and machine learning recognized this early on, and modern techniques employ far more sophisticated strategies for making changes, large and small.

Hill Climbing has the some of the same problems as Generate & Test. They're pretty blind, stupid strategies. Both can tell you what stands a better chance of working or not in particular situations if you don't have much else to go on. But they can't tell you *why* and what to do about it if the results aren't satisfactory. They don't have any theories.

Now we've got our answer as to why it's so hard to deploy innovation in today's society. Innovation isn't about random (generated) ideas or bit-by-bit change. An innovation usually offers a whole new *theory* for why things are the way they are, and requires a radically different approach. If the only mechanisms our society offers for adopting new innovations are Generate & Test, and Hill Climbing, it will be hard to adopt innovation.

What we're doing in this book is giving you a new theory. Up to you to decide if it makes sense.

Goal stacks

We've already talked about the idea of a *goal stack* (in the ...*Even Possible* chapter [Ch. 4]), solving a big problem by breaking it down into smaller and smaller *subgoals*, and finally down to specific actions. This is the "divide and conquer" heuristic that people commonly use in everyday life.

In political and economic hierarchies, this breakdown is embodied in people. Leaders are tasked with planning to solve ambitious goals by deciding what the subgoals are. Lower-level employees are tasked with carrying them out by taking small steps. Orders flow from top to bottom, almost never in the other direction. If everything goes according to plan, this usually works.

But in the real world, everything *doesn't* always go according to plan. Is the subgoal the right way of achieving the larger goal? Did each action have the result that was intended? If not, you can't just keep working on the current goal. You typically need to *go back up* the goal stack, and reconsider higher-level goals. Maybe you need to *replan*, abandon the current plan and break up the high level goal into a completely different set of subgoals. AI has a subfield, called *partial-order planning*, that deals with these issues.

That's what innovation is. Innovation proposes a new solution to a high level goal, that obsoletes an existing subgoal plan. In effect, its "*redivide and conquer*".

For example, in transportation, we can adopt the goal of trying to make a more fuel-efficient car, or a train that goes faster. Those would be incremental advances to already-existing plans. But we could also propose an innovative new transportation system like PRT (see the *Transportation* chapter [Ch. 29]), that is neither a car nor a conventional train, but does solve the higher-level problem of urban transportation, which is what we were trying to do in the first place.

In personnel hierarchies, the problem is that people are assigned and committed to implementing incremental plans. Adoption of the innovation is a threat to them. The best chance for innovation is in those rare cases when the goal itself is entirely new, or there is no other existing plan to achieve an existing goal.

Low-level people do not have the authority to go up to a higher level and change the plan. High-level people are also disinclined to adopt innovation, because it challenges their competence in executing the original plan obsoleted by the innovation. They are already highly invested in the current plan, and they perceive the innovation as being too risky. Often, they incur some costs and risks in implementing the innovation, and reap no immediate, personal benefit if the innovation succeeds.

It is the nature of the hierarchy itself that presents the major obstacle to innovation. We elaborate this argument in the chapter *No Leaders* [Ch. 22].

Who decides about deploying innovation?

Let's review some of the players.

Big government

US Democracy is supposed to be working for the people, but in *The Trouble with Voting* [Ch. 23] we argue that it is an institution supported primarily by large corporations to help them efficiently preserve the status quo. It *precludes* innovation.

True, governments fund research, though investment in research is a trickle compared to other major expenditures. But the tax code, our laws, enforcement, our process for selecting leaders, and increasing income disparity, all support the view that government is more interested in the status quo than innovation.

One of the most important kinds of innovation we can have is to drastically reduce the cost of a product. Since most of the cost of a product is in labor, decreasing production costs usually means fewer jobs required to fulfill demand. To a politician, jobs means votes. To a labor leader, jobs mean union membership. To a contractor, jobs mean billable hours.

There are many ways that leaders preclude change. You'll hear:

"We need another study to make sure its safe."

"We don't want to gamble with the taxpayers money".

"We're putting out a request for proposals to make sure we get the best deal."

Many boil down to delaying tactics and these can literally go on forever. This is why we need complexity management tools (see *Tools for Reasonocracy* [Ch. 27]).

Little government

"Power tends to corrupt, and absolute power corrupts absolutely". So it stands to reason that smaller governments, i.e. local governments, should be less corrupt. However, we see a trend in national parties gaining increasing influence in local government. For instance, on local labor union laws, and on gerrymandering congressional districts [Fang 2014].

Big companies

Companies will tell you they're always open to innovation. "Build a better mousetrap and the world will beat a path your door." Rarely does it happen. But it does happen.

More often, if they can't defeat a new idea in the marketplace, big companies will "buy it and kill it". Eventually it becomes too widespread to "buy and kill", so some company will "buy and promote". This is a much slower process than need be, but even the most entrenched status quo gives way eventually. "The average life expectancy of a multinational corporation-Fortune 500 or its equivalent-is between 40 and 50 years." [Foster 2015].

Little companies

If big companies are hamstrung, startups and small companies are supposed to be the vehicle for innovation. But the survival pressure on small companies is such that they are often tempted to stop starving to death and promote an incremental solution, even if it is insufficient for the customers' needs. Companies want innovation—but *not* primarily to help their customers. Rather, it's to get *competitive advantage* over other companies. Even a little advantage is enough to defeat another company—why risk more?

If an incremental improvement is successful, it *fills the niche* of "some kind of improvement". Because of the overhead of convincing the powers-that-be to consider any new solution at all, it may then become harder to consider a more radical solution that might actually be sufficient.

Capitalism enforces this process. Proposers of the incremental improvement become another obstacle that our radical innovator has to overcome.

Example: You might think the fuel efficiency of cars is improving. [WantToKnow 2017] says the 1908 Model T got 25MPG and the 2013 cars average 24.7MPG. Those 2013 cars are way better than the Model T in most respects, but fuel economy doesn't happen to be one of them. We have "filled the niche" of "better car" with more cupholders, air conditioning, reliability, etc. but that's not helping carbon ppm. (Electric cars are promising to break this dismal trend.)

Here's another, all too common, process under Capitalism. Two inventors have different, insufficient solutions to a problem. They work at different companies. Perhaps these two solutions don't preclude one another, they're merely partial solutions. Capitalism discourages them from teaming up and making a sufficient solution due to the competition between their respective companies. *"Divide and be defeated."*

Consumer reluctance

Companies say that competition between them drives innovation. But, in the *Can Capitalism be saved?* chapter we show that much of the *real* competition is between a company and its customers.

The company says "Customers are our #1 priority". This is almost never the case. Making a profit is. They make a profit by getting customers to pay as much as possible for a product. To do that, marketing, to be polite, frames the product in the best possible light. To be impolite, they lie. By doing so, each such company decreases the trust a consumer has in companies in general.

So when a new (or old) company comes out with what they say is an "innovation", customers are more than a little wary. This gives innovation a bad reputation, and prevents consumers from taking the risk of adopting innovation.

Is that mousetrap really better?

Of course, not every proposed innovation is really worth adopting. Before adopting an innovation, we should really do *cost-benefit analysis* to understand whether the benefits outweigh the costs and the risks.

Often overlooked, also, is the *opportunity cost*, that is, the cost of doing nothing or failing to adopt complete solutions. When Republicans claim that climate measures will "hurt the economy" they are perhaps correct in the short term but incorrect in the long term.

Some mousetraps aren't better

The vast majority of inventions fail. They are, by definition, unproven and therefore risky. Often the "innovation" is good for some particular reason or situation, but other reasons or broader situations make it not worth it. We can't eliminate this risk, but we can be strategic with analysis to kill bad ideas before they waste too many resources. There are often "plateaus" in design spaces. Given a set of parameters, we can often preclude an infinite design process with the rationale of: "It's not likely we can do much better".

Bang per buck

Bang-per-buck is a useful metric for comparing solutions, especially when there are various quantities of something that you can buy. Suppose product X is cheaper than product Y but it is less effective than product Y. We can buy more of X to make up for its less effectiveness. If X is better bang-per-buck wise, X is a better solution than Y.

Net solutions

Here, we don't mean "net" as in Internet, it's "net" as in "after everything is considered". Any complex solution is very likely to have both good and bad in it. For instance: a refrigerator that uses little energy to run, may take so much energy to produce (embodied energy) that over its lifetime, it is no better than a less efficient-to-operate refrigerator.

The same kind of thing can happen with money. Using an efficient product saves on operating expenses, but the up-front costs are so high they will never be recouped by the 10 year lifetime of the product.

Evaluate a solution on its net benefit, not the benefits of a given feature in isolation.

Partial vs precluding solutions

Say our goal is an "off-the-grid" house, one that doesn't require electricity from a citywide source. There is a refrigerator that uses half the energy of our old one. It will not get us off the grid by itself, so it is an insufficient solution for our goal, but it is a *partial* solution. The efficient refrigerator in combination with other efficient appliances and solar panels complete a sufficient solution.

Let's say there's a new maximally efficient inverter for our solar energy system. It produces 37 volts which is incompatible with all appliances. Using it would preclude the other parts needed to make it a complete solution event though the inverter is more efficient. Hill climbing gets us the more efficient inverter, but it strands us on a foothill, wherein the real mountain we want to climb is the "whole house efficiency".

Sufficiency of solutions

A very important consideration is whether the proposed solution is *actually sufficient* to solve the intended problem. A big problem with incremental solutions is that people will get distracted by the fact that they promise an improvement. They might then ignore the fact that that improvement may not lead to a sufficient solution. Fry's adage: *"All that's necessary for the bad guys to win, is that the good guys are distracted by the insufficient."*

Nowhere is this more evident than in proposals for dealing with climate change. Consider the amount of the carbon in the atmosphere, now 400 parts per million (ppm). The sustainable ppm is 350. [350.org 2017]. It's rising at 2 ppm per year. Suppose we have an array of proposals:

Proposal	Result
0. Do nothing	Unchecked emissions. Disaster!
-1. Better fuel economy	Reduced acceleration of emissions
-2. Paris Climate Agreement	Limited emissions growth
-3. No fossil fuels. PRT.	Slow decline in emissions.
-5. No emissions. Sequestration.	Radical decline in emissions.

Don't take this too literally—in this chapter we're trying to make a general point about solution methods, not the details of various climate plans per se (but see *Transportation* [Ch. 29], about Personal Rapid Transit (PRT)).

Solution 0, doing nothing, is roughly what we're doing now. That causes the situation to worsen by 2ppm per year. Not a good option if you're interested in long-term viability of the planet, but that's the overall "plan" of US Democracy, the world's worst polluter per capita. We humans tend to delay hard choices until true catastrophe hits.

Solutions -1 and -2 are "politically achievable", and make activists feel good because they are "improving the situation". But they are insufficient *niche fillers*. They won't tip the balance to reducing atmospheric concentration of carbon, no matter how successful they are at getting adopted.

Solutions -3 and -5 might be "politically unacceptable", but they'd be *sufficient*. They would likely constitute a means of reducing carbon ppm and halting climate change. Some methods of sequestration, such as geoengineering, might involve a lot of risk, and it's not clear we'd have to go that far, but it would be worth considering.

Solution -5 is the least likely to be chosen even though it could theoretically get us to a healthy state faster. If we were deciding between -3 and -5, we'd hear "The perfect is the enemy of the good." and that argument would have merit. But, applied to comparing -1 to -3, it would be incorrect. -3 is more perfect than -1, and riskier to deploy, but -1 is just plain old insufficient.

Search by design

With our complex world, how can you find the best that's out there? Even with today's advanced web search engines, you have to know the right words to enter to search for.

One strategy is to *search by design*. First design what you think the ideal solution would look like, not in detail but enough to clearly articulate high level product features.

Say you are looking for efficient cars in the USA. If you understand car usage, you know that average people in a moving car is about 1.2. So your "ideal design" might have 2 seats. If they have any more, that's extra weight and volume that wastes energy. Once you realize that, it becomes easy to recognize inefficient cars—they have more than 2 seats!

Most products will have numerous criteria. Search first by the most restrictive one, and filter the remainder with the other criteria.

Recap: The real Innovator's Dilemma

The creative spark of innovators is alive and well all over the planet. This chapter argues that the hard part of innovation is moving that "a-ha" from idea to reality at scale. Clayton Christensen's *Innovator's Dilemma* [Christensen 1997] posits that companies favor incremental changes over disruptive ones, even though that strategy may kill the company. The memes of "Too little, too late" and "The road to hell is paved with good intentions" surround insufficient solutions.

Nobody will admit to being anti-innovation. But most big decision makers act that way. They *have to*—they are embedded in command-and-control hierarchies, where they have incentive only for incremental change and not innovation. They work in business and government structures that have only Generate & Test and Hill Climbing as improvement heuristics.

Science doesn't limit itself that way. The reason that science is so much better at generating and adopting innovation is that it is willing to go up its own goal stack when it needs to. It considers not just incremental improvement, but *theories* about why things are the way they are, and what they could be. If a new theory requires non-incremental change, well, so be it. Redivide and conquer is embraced, not excluded. We can look particularly to the science of AI for inspiration, which explicitly studies goals, planning, action and change.

In *The Structure of Scientific Revolutions* [Kuhn 1962], Thomas Kuhn dissects how this process works (and the obstacles that science, too, faces, when an innovation is too disruptive). This is why we advocating making economic and political structures work more like the social processes of the scientific community (see *The process of Science* [Ch. 24] and subsequent chapters). Maybe then we'll get our better mousetraps.

Afterword

It's easy to get discouraged when you see the state of the world today. The technologies we advocate are presently being used by governments, militaries, and corporations as a weapon for their competitive ends. What would need to change, for technology to be redirected towards the ideal of "getting along"?

That's of course the big question. It's not surprising that capitalists and militarists in today's society are using whatever technology they can get their hands on. And we agree with the pessimists, that if technology only serves selfish purposes for an elite, we're headed straight for a collapse scenario.

But it doesn't have to be that way. Capitalists have to sell their system to the pubic to get the public to cooperate with them as workers and customers. Militarists have to sell their system to the public to get them to cooperate as soldiers, voters and taxpayers.

They do this by hoodwinking people into believing in the myth of *inevitability*. Margaret Thatcher was fond of promoting the idea that, after the fall of Communism, neoliberal economics and parliamentary democracy were the only viable systems. She used the acronym *TINA: There is No Alternative.*

She was wrong. For example, in the one-shot Prisoner's Dilemma, the suspects "inevitably" choose defection. But we know now that this inevitability is indeed avoidable, if you go meta and look at the larger situation. That's what we're asking you to do—think outside the box of conventional politics and economics.

Capitalists have to convince people that competition, wage slavery, economic inequality, etc. are absolutely necessary to achieve adequate productivity. Militarists have to convince people that war is inevitable. That you can't possibly be safe without an army. Politicians have to convince voters that people who speak a different language, have a different skin color, religion, political system, dress differently, etc. etc. are your "enemy". (We particularly dislike that pejorative word, *enemy*. It's a label that's used to dehumanize people who have differences.).

What we *can* do is show people that the myths of inevitability are false. We first have to get people to believe that there are viable alternatives. We can then work on choosing the best of these alternatives.

Automation *can* mean productivity without requiring wage slavery. Sustainable economic production and sensible consumption *can* get us out of "competing with nature" until we experience environmental collapse. Producer and consumer cooperatives *can* organize people without exploitation or hierarchy. Technology *can* allow producers and consumers to collaborate, where the size of such organizations range from one to millions.

Consensus decision-making, Reasonocracy, and better education *can* replace authoritarian regimes and political fighting in government. Cooperation between neighboring peoples actually keeps people safer than any military could. When was the last time there was a war between Massachusetts and Connecticut?

Pragmatists ask, "if your solutions are so good, how come people aren't already doing them?". We're familiar with this objection from our careers in scientific research. Whenever somebody proposes a new research idea for which all the required elements are already in place, it's easy to imagine that there must be something wrong with it. Otherwise, it would have already been adopted. People chronically *underestimate* the difficulty of making large-scale social improvements, and *overestimate* the difficulty of making technological advances.

The truth is that in almost every field, the design space is so big that most possible choices simply haven't been explored yet. The new idea seems obvious, but only in retrospect. If you assume an idea has already been tried and failed, you would reject all new ideas. Good ideas have to start somewhere.

Another way to look at it is best expressed by a quote from science fiction writer William Gibson: "The future is already here, it's just not evenly distributed". Throughout the book, where possible, we've pointed to present-day, small-scale examples of directions we advocate: cooperative social organization, the maker movement, and information technologies. The challenge now is to learn the lessons from existing examples, and apply the principles to solve our large-scale problems.

Our best guess is that so many people believe our big problems are so inevitable, that they won't consider alternatives. The technologies of Makerism, AI, and the others we present (and even the math of the Prisoner's Dilemma) are so new that people don't fully understand the consequences. The route to the cure starts with education. In this book, we're trying to do our part. Maybe we can all just get along.

Appendices

Appendix 1
Essential reading

This book has a lot of references. We want to acknowledge the many influences that led us to write this book, back up our facts, and give you resources for continuing to think about the issues we raise here. But we didn't want to have 100 pages of references, especially in printed books.

Online readers should just be able to click the citation in square brackets. Print readers can look up the reference at our site, http://www. whycantwe.org.

In this chapter, we list just a few of the most important book-length works that we consider essential reading. Now that you've finished our book, make some of these be next on your reading list.

Part 1: What keeps us from getting along?

[Axelrod 1984] Robert Axelrod, *The Evolution of Cooperation*, Basic Books, 1984.
The classic work on the Prisoner's Dilemma. Describes the simulation experiments, and many different possible strategies. Essential reading on our central concept. More about the math itself, than its application to societal problems. Implications for society are what we do here in our book.

[Wright 2000] Robert Wright, *Nonzero: The Logic of Human Destiny*, Vintage Books, 2000.
Makes the argument that evolution selects for positive-sum games. The best book on evolution since Darwin. The subtitle is not an exaggeration. Eye-opening.

[Mullainathan and Shafir 2013] Sendhil Mullainathan and Eldar Shafir, *Scarcity: Why Having Too Little Means So Much*, Times Books, 2013.

Explores the psychology of scarcity, its pros (!) and cons. The first sentence: "We were too busy not to write this book". For ourselves, we can say, "We were too hopeful for the future and too scared of the present not to write this book".

It's why we believe that conquering scarcity by technical means holds the key to transforming society, far beyond the solely economic impact.

Connecting the dots between these three books (and the Kohn below) is what got us the central thesis of our book.

Part 2: Does human nature allow us to get along?

[Rosenberg 2003] Marshall Rosenberg, *Non-Violent Communication: A Language of Life*, Puddle Dancer Press, 2003.
Gives practical advice on how to communicate in an empathetic and cooperative way. Shows how people get stuck in traps that lead to arguments, and how to avoid or get out of these traps. Gives us hope that we can improve interpersonal communication on a one-to-one basis, a prerequisite for building a more cooperative society for everyone.

[Susskind, McKernen, Thomas-Lamar 1999] Lawrence Susskind, Sarah McKearnen, Jennifer Thomas-Lamar, eds. *The Consensus Building Handbook: A Comprehensive Guide to Reaching Agreement*, Sage Press, 1999.
A thick book of practical ideas for running meetings, negotiations, and any kind of group cooperation. Emphasizes cooperation on shared values rather than "hammering out differences". Susskind was a founder of the Harvard Negotiation Project, which also produces the excellent *Getting to Yes* series of books, available at any airport bookstore near you.

[Kohn 1993] Alfie Kohn, *Punished by Rewards: The Trouble with Gold Stars, Incentive Plans, A's, Praise, and Other Bribes*, Houghton Mifflin, 1993.
Explains the important difference between *intrinsic* and *extrinsic motivation*. Kohn is a philosopher of education, and writes about schools. But this is a key point that goes way beyond education. Forcing or bribing people to do things is inherently oppressive. Freedom and self-actualization depend upon intrinsic motivation.

[Kohn 1986] Alfie Kohn, *No Contest: The Case Against Competition*, **Houghton-Mifflin, 1986.**

He makes the case against competition so compellingly, we don't have to (but we did). "The main thing public education teaches you is how to compete." Again, he's writing about education, but the arguments apply to competition in economics and politics equally well. Convinced us that over-reliance upon competition is the root cause of most of society's problems.

Part 3: Can we get along economically?

[Lipson and Kurman 2013] Hod Lipson and Melba Kurman,
Fabricated: The New World of 3D Printing, **Wiley, 2013.**
The best book on the transformative technology of 3D Printing, the hardware that will enable Makerism. This is moving so fast that you probably want to make sure you read the latest edition.

[Drexler 2006] K. Eric Drexler, *Engines of Creation: The Coming Era of Nanotechnology*, **Doubleday, 2nd edition, 2007 (1st edition 1986).**
Another "make anything" technology is *nanotechnology*, essentially a 3D printer at the microscopic level. True nanotechnology appears more difficult at the moment than the advanced 3D printers we describe, but ultimately the two technologies are synergistic. Three decades later, *Engines of Creation* remains a seminal work for both technology and its social consequences, all in an easy to open package.

[Minsky 1986] Marvin Minsky, *The Society of Mind*, **Simon and Schuster, 1986.**
Conventional psychology is wandering around in the dark. Marvin turns on the lights by *modeling* the human mind in a way that can plausibly be built. If you're curious about how you think (and who isn't?), read *Society of Mind* and its 2006 sequel, *The Emotion Machine*.

Oddly enough, we had trouble figuring out a single book to recommend on AI for beginners, that would have the proper perspective for this book. The standard textbook on AI is Stuart Russell and Peter Norvig's *AI: A Modern Approach*, [Russell 2009] but it's for technical readers only. A recent just-the-facts overview from a blue-ribbon commission (if a little dry and conservative) is: http://ai100.stanford.edu [Grosz 2016].

Part 4: Can government help us get along?

[Madison, Jefferson, et al 1787] *The Constitution of the United States of America*, Constitutional Convention, Philadelphia, 1787.
A colossal achievement for its time, but like just about every other invention from over two centuries ago, it's past its expiration date.

[Kuhn 1962] Thomas Kuhn, The Structure of Scientific Revolutions, University of Chicago Press, 1962.
Insight into the social processes of science. Shows how "revolutions" can take place without anybody getting hurt. Unlike political revolutions.

Part 5: Can we get along in...

[Swanson 2010] David Swanson, *War is a Lie*, Just World Books, 2010.
The definitive refutation of every justification you've ever heard for participating in a war.

Appendix 2
Frequently Asked Questions

I'm skeptical of this whole thing

Q: I'm intrigued by the book title. So tell me, in a nutshell, why do you think we haven't been able to get along? What can we do about it?

A: Many readers are so curious about our answer to the title question, they can't wait to plow through the book to hear it. So here's the short answer: Basically, it's because many people misunderstand the tradeoff between cooperation and competition.

People often think they should compete in a given situation, when in fact, it often makes more sense to cooperate. And it's the excesses of an extremely competitive attitude fomented by our divisive culture, that make people aggressive. This causes war, poverty, and a host of other ills. Technological changes are rapidly decreasing the advantages of competition and rapidly increasing the advantages of cooperation.

While Rodney King's original may have been a rhetorical question, the real question is precisely "how can we get along?", given the societal pressures that derail cooperation. This book details why it is hard, and proposes some new solutions. Read the book for the whole story.

Q: This math stuff is clever, but it doesn't describe all the complexities of the real world.

A: It's true that the math we present is an idealized abstraction, and people criticize us for leaving out many cultural and historical factors. These factors also contribute to the problems we describe, or block our proposed solutions.

But science always makes progress by thinking about simplified models, then if necessary, adding the complexities back in later. If you consider all the complexities at once, you get paralyzed. And paralyzed we are, when it comes to politics and economics.

Understand our argument about math, evolution, psychology, and technology, first. Don't reject it out of hand because we didn't consider everything. As physicists say, "Physics is the science of frictionless elephants".

Q: You want to change so much of today's society. How we gonna get from here to there?

A: That's the part we're least sure about. What we're trying to do in the book is to give you our vision of a more cooperative society. For the moment, put aside your concerns about the difficulties of getting there, and just see if you agree with our vision and our analysis. We do present some strategies and intermediate steps to demonstrate that it is possible to drastically improve our condition, or at the very least, to show that there are some plausible, untried ways of achieving it. Once you understand where we're going, we invite you to join us in strategizing about how to get there.

Q. By criticizing the status quo, aren't you biting the hand that feeds you?

A: Guilty as charged. We're a couple of healthy, white, male baby boomers born in the USA's most prosperous urban areas (NYC and LA) into caring, intelligent families. We've had relatively excellent educations and traveled extensively in and out of the USA. We live in one of the world's most intellectual cities (Boston) and have worked in and around the one of the most advanced technical universities (MIT). We had early access to the Internet and the best of computing resources, which we helped develop.

But the vast majority haven't been so lucky. If we are to maintain our high standard of living and include the rest of humanity in it, let alone advance it, the architecture of this pyramid has to change. This book aims to foster the good will in most of us, and innovation in the most creative of us, to accomplish that.

Q: What if there are unintended consequences?

A: The foreseen consequences of the status quo are pretty bad. So we had better do something new. We wrote this book largely to foresee the consequences of enacting our proposed solutions, as compared to continuing on our present path. We spend a lot of the book trying to improve decision-making processes, which should help us foresee, and also deal with, unintended consequences. New tools can help us augment our species already formidable creative powers. See the chapters on *Reasonocracy* [Ch. 25] and *Deploying Innovation* [Ch. 34].

Q: If your ideas are so great, how come they haven't already been implemented?

A: We hope that we're bringing a new perspective, with our arguments from mathematics, evolution, psychology, artificial intelligence, and 3D Printing. Let us know if we succeeded.

Some of our proposed solutions aren't considered, or adopted, because there are vested interests fighting to maintain the status quo. See our answer below to the question, "Won't the status quo fight this?". See our analysis in *Deploying Innovation* [Ch. 34], and our final encouraging words in the *Afterword*.

From our research careers, we're familiar with objections of the form, "There must be something wrong with the idea, otherwise it would have already been done." Think that way, and you'll reject any possible innovation, out of hand.

You're way too optimistic about human nature

Q: Conflict and competition are inevitable. You'll never get everybody to cooperate on everything...

A: Well, we're not always going to get everybody to cooperate, no matter what, nor should they in every case. But there's a lot of needless conflict and competition going on in the world. That's what causes war, poverty, and a host of other societal ills. What do we do about it?

We have to figure out when competition actually necessary, and when is it just due to people's fears. In *Jailbreaking the Prisoner's Dilemma* [Ch. 1] and *Survival of the most cooperative* [Ch. 3], we present some tools for thinking about how to figure out in what cases it makes sense to compete vs. cooperate.

We particularly dislike the stance that conflict or other societal problems are "inevitable". If you think something's inevitable, you'll give up on trying to fix it. Besides, few things are literally inevitable.

Q: How do we resolve disagreement?

A: With Makerism, agreement is much less necessary because trade is much less necessary because you make what you need. Structural reforms of government can get better motivated people utilizing more reasonable processes to make better decisions (See *Introduction to Reasonocracy* [Ch. 25].)

Q: War is inevitable. We've got 10,000 years of history to prove it.

A : That was then, this is now. Humanity has managed to (almost) put an end to human sacrifice, cannibalism, slavery, and a host of other barbaric practices that had gone on for thousands of years prior to their end. Why not end war? See the chapters *War* [Ch. 33], *War crimes* [Ch. 32].

Q: Will we need a military ?

No. Since wars and the militaries that make them possible are unreasonable, our proposals for a more rational government will mean the end of militaries as we know them. Wars now occur because they benefit numerous powerful actors, including: Presidents, an elected Congress, misguided voters, corporate executives, poor people needing a job, bored patriots, and psychopaths in power. Our proposals show how to eliminate the above roles. We hope that the primary role of much smaller armed services will change from enforcing political power, to disaster relief and humanitarian efforts.

Q: We don't need new tech, we just need to be kinder to each other.

A: We sure do. But right now, the pressure of our competitive society gives people lots of reasons not to be kind to each other, like poverty, competition for status, and fear of various threats. Technology can save most of the labor now spent creating wealth.

Without having to have a conventional "job", we can afford to be kinder.

We agree that there are many steps that can be taken towards a more cooperative and empathetic society that don't involve technology. Social and ethical education (See *Constructionism: Education for Makerism* [Ch.

28]), psychotherapy and counseling, meditation and (some) religious practice, reducing the sexist macho culture that encourages aggression, etc. can also help.

Q: What about evil people?

Eliminating scarcity will reduce the motivation for many ill-intentioned people, who rationalize their selfish behavior by their perceived need for resources. Society now selects for aggressive people with psychopathic tendencies, and elevates them to "leadership" roles (See the *No Leaders* chapter [Ch. 22]). Without conventional "jobs", people won't be at the mercy of psychopathic bosses and bosses won't have assistants to help them.

Changing our legal system into a justice system (see the *Justice* chapter [Ch. 30]) should drive fewer people crazy. Teaching cooperation should help to make a more empathetic, less confrontational society. Advances in psychology will provide more insight into why people with various mental disorders commit crimes, and how to deal with them.

Ending evil is probably an unobtainable goal. Reducing it to a fraction of its current presence is feasible.

I'm not really a technical person, so this doesn't look so great to me

Q: Techno-utopias have been promised for decades. None have worked on a significant scale. What makes you think this time it is any different?

A: First we agree that a myriad of schemes have failed. But let's point out some successes of technology. Cars are far faster than horses. The age-old dream of flying is now not only common place, but can be two orders of magnitude faster than birds. The Dick Tracy two way radio worn on a wrist proposed in the 1950's has been surpassed by cell phones and smart watches. The web not only achieves fast, smart communications all over the planet but is deployed to billions of people at low cost. These technological developments have led to social progress in improvements of standard of living, and fewer wars, as Steven Pinker points out [Pinker 2011].

But there's been three key pieces missing .... until now. The first is an understanding of the mathematics of cooperation due to theoretical work on the Prisoner's Dilemma. The second is the technology of AI and personal manufacturing, which holds the promise of solving scarcity. The third is to develop social processes for making consensus-based decisions that will replace our current power-based governance. We present such a proposal in *Reasonocracy* [Ch. 25].

Q: What if I don't want all this new-fangled techology?

A: We believe that, given choice, the majority of people will choose technological progress over static, tribal societies. They have for centuries. And even if you don't want to personally use high-tech, it will still benefit you.

For the minority that prefer less technology-intensive lifestyles, they should form intentional communities that implement these lifestyles. Note that we are in favor of intentional communities in general as vehicles for lifestyle experimentation. Care must be taken to make sure that these communities aren't steamrolled by outside competitive forces, as they often are now. A positive example: the Amish. A negative one: Native American "Indian Reservations".

Q: You scientists think everything's a technical problem. It's really a people problem...

A: We need both technical and social solutions. One side without the other isn't going to cut it. One of the reasons the big problems are so hard is that it hasn't been easy for technical and political people to work together. Technology increases understanding, raises standards of living, and relieves pressure that distorts social relations. Anything we can do on the social side to improve social relations, and help people feel better about themselves and others, and act positively, is great.

We confess that, as technologists, we have more to contribute on the technical side, so it's emphasized in this book. But our message to people in politics, business, psychology and the social sciences is that thinking about the scientific issues we present can help them achieve their goals of a more cooperative and humane society. Political "activism" alone, or self-help psychology alone, won't bring about enough change, if it's still embedded in a society mired in scarcity and competition.

Q: Is there a danger that robots will go berserk and kill people like in the science fiction movies?

A: Unlikely. The problem with all the Frankenstein-like scenarios is that they posit technological progress sufficient to create an intelligent robot, but they don't foresee any progress in social and emotional intelligence, which has admittedly been slow. What AI research is now discovering, is that social and emotional understanding is an essential part of what it means to be intelligent.

We're hopeful that by the time we get around to having human-level AI, we'll have figured out how to have programmed in ethical behavior. AI's won't act like a James Bond villain; they'll be smarter than that. See the chapter *AI: Not the son of Frankenstein* [Ch. 17].

Q: Won't a society highly dependent upon robots and machines be dehumanizing?

A. Do you find your car (a transportation robot) dehumanizing? How about your refrigerator? A lawn mower? An electric drill?

Most of the downsides that Luddites or technophobic leftists associate with adoption of technology: over-commercialized culture; externalities like pollution; "dehumanization"; privacy violation, suppression of alternative cultures; imperialism, etc. are actually pathologies of the scarcity/competitive/Capitalist culture. They are not inherent to technology.

Because Maker tech (see the *Makerism* chapter [Ch. 15]) is so cheap and under the control of individuals, it won't share the dehumanizing nature of large-scale industrial technology. Because Capitalism depends on economies of scale, large numbers of employees and large numbers of consumers all have to behave identically for it to work. That's what causes the rigid, dehumanizing forces.

Q: Maybe OK for the third world, but here in the first world we've already got it pretty good. You want risk disrupting all of that?

A. We certainly do have it pretty good in the first world compared to other places. But the first world is disrupting itself. Problems like climate change, growing inequality, and unsustainable consumption, left unchecked, may threaten our quality of life in the future. And third-world problems cannot be walled off from affecting the first world.

Disease, immigration forced by economic necessity, terrorism, and other problems are now spreading from the third world to the first. We need to think broadly about problems faced by everybody, not just our own situation.

Q: Maybe OK for the first world, but what about the chaotic third world and all its diverse cultures?

A: Some technological advances and social innovations may be harder to accomplish in the third world, for those reasons. But the third world actually has some advantages as a venue for change.

The need is greater, so people may be more motivated to try out-of-the-box solutions. Smart people in the third world may realize that they don't have to repeat all the mistakes made historically by first-world countries. Third world countries are now moving from agricultural societies directly to the Information Age without going through the mechanical, industrial era that caused so much heartache. Furthermore, cultural diversity can be an asset, leading people to imagine innovative or culturally appropriate solutions that couldn't have been thought of in an already-built-out society.

I don't believe this 3D printer stuff is as great as you say it is

Q: Where do the raw materials come from for the printers?

A: Plastic can be recycled from water bottles and unused plastic objects into filament for printing. The core material of some "bioplastics" is corn starch. Algae can be grown in microfarms, from which oil can be extracted, forming the basis of plastic that can be printed. Cellulose and hemp are other fast-growing plant material useful for raw materials. Sand can be melted by focusing the sun to form glass. Combining sand and impurities can make many ceramics. Aluminum and iron can be separated from dirt. Carbon is plentiful and can be formed into nanotubes or sheets of graphene to make strong and highly conductive material. Alternative designs can take advantage of local materials for making functional equivalents.

Q: Without electrical grids, how do you supply energy?

A: An advanced 3D printer (not yet invented) will be able to print solar cells and batteries. Wind and water can also create decentralized electricity.

Q: 3D printers sound too complex to be reliable.

A: As complexity goes up, reliability does go down, but modern integrated circuits have billions of parts (transistors) in them, have clock cycles of billions per second, and are quite reliable. (You couldn't read this if they weren't.) Redundancy and other fail-safe mechanisms can help with reliability. Designing for repairability and on-line resources like Repair Clinic (http://www.repairclinic.com) will help.

Q: This home 3D printer sounds too complicated to learn how to use.

A: People can drive cars, operate computers and smart phones, all pretty complex machinery. In the early days of automobiles, drivers needed to be competent mechanics, but that's no longer true. We will be able to make even better user interfaces than we have now using natural language processing and augmented reality. We expect people to spend more time making, but that will be more than compensated for by not having to shop, manage money, find and work at jobs you don't like.

Q: Sure, you can make some kind of food artificially, but won't it be like Velveeta replacing my organic goat cheese?

A: Matching food perfectly to the smell, taste and texture of something you're use to may be difficult. Making it nutritious and taste good is more doable. We expect many makers to tune and share recipes. With the multitude of options available, numerous will be rated by the adventurous, and the timid can choose amongst them.

Q: Sure, the 3D printer can print some kind of lamp, but won't it be cheap plastic junk rather than my hand-carved rosewood lamp?

A: 3D printer filaments containing wood already exist. [Matterhackers 2017] lists 53 filament types including wood, metal, ceramic, silk. These materials won't be identical to their natural counterparts in all respects, but the list of printable materials is expanding rapidly. From the description of the Cherry Wood filament at this website we quote: "Rough or smooth surface possible during one print. Paintable, grindable, carvable and stainable. Printable tree-rings"[Brewster 2014].

Q: Can a printer make very strong metal objects?

A: [Simon 2015a] and http://auroralabs3d.com describe a printer that makes stainless steel and titanium objects. This printer has a resolution of about 100 microns. It uses metal powder sintering technology. The initial version is $33K. https://markforged.com can print carbon fiber parts that can be as strong as aluminum. Graphene is a material that is a hundred times stronger than steel per volume, and conducts electricity with half the resistance of copper, and even less than silver. It also has superior thermal conductivity. It is being used in advanced 3D printers [Benchoff 2015a].

Q: Which century will this scarcity reducing technology be invented, and which century will it be widely deployed?

A: We admit to being poor at estimating the arrival of advanced 3D printers that are able to make all of their own parts. 3D printers are already capable at making many useful things and progress is rapid. In Sept 2015, a version of Reprap (Snappy) printed 73% of its parts [Benchoff 2015b]. 2016 had 185 Maker Faires in 32 countries. The September 2015 NYC one had more than 90K attendees with more than 900 exhibits by makers. See *A day in the post-scarcity life* [Ch. 18] if you're still skeptical.

Q: We already have an unsustainable ecology with our current standard of living but if we raise the overall standard of living our planet will be even less sustainable.

A: Our use of resources is inefficient to put it mildly. There is no physical reason why we can't increase standards of living world-wide and, at the same time, decrease our ecological impact. Appropriate technologies can be better in all important aspects than what they replace. It will take technological innovation, advances in government, legal and educational systems. This book describes the path.

Q: Printers will be used to print dangerous things like guns and drugs, creating big problems.

Without advanced printers, guns and drugs are already widespread. Plus any flexible tech can be used for ill, as 9/11 showed with airplanes. By solving scarcity, we get rid of the need to protect property with a gun and the need to make money pushing drugs. We can print houses for the homeless and you won't have to waste money on insurance. We'll also have more resources for tackling other problems.

Q: Won't getting everyone an advanced printer be too expensive?

A: Part of our definition of advanced printer is something that can make all of its own parts. It will make the solar cells and batteries it needs for energy as well as machines for gathering raw materials. If a friend doesn't have a printer and you do, you'll print one out for him (or at worst, the easy-to-assemble pieces), as he'll do for his other friends. Printers will be free like most other material things you'll need.

You think you can replace Capitalism? Dream on, kid.

Q: If we remove competition in the economy, people will just invent new things to compete about, like social status.

A: Could happen. But it doesn't have to be like that. Once people can make what they want, we suspect that social status won't be as tied to material wealth as it is now. And you have to wonder about people whose desire for wealth or status is motivated by the need to feel superior to other people. Why are they so invested in feeling superior to others? Where does that come from? It can't be psychologically healthy to be that way.

Like a lot of other issues in the book, it depends on whether you're fundamentally optimistic or pessimistic about human nature. As we confessed at the outset, we're optimists. If you think people are so inherently aggressive that they'll make any excuse to fight with others even in the absence of real need for conflict, you're ignoring evidence to the contrary. See the section *Keeping up with the post-scarcity Joneses* in the *Interpersonal relations* [Ch. 5] chapter.

Q: There will always be scarcity, because no matter how much people have, they'll always want more. (This is called Jevon's Paradox.)

A: Again, maybe, but we think it won't turn out that way. People don't "always want more", though Capitalism certainly encourages this attitude. In reality, diminishing returns always set in. After a certain point (and that point is surprisingly low), studies show that additional wealth doesn't make people any happier.

People whose greed is insatiable have a mental illness akin to Obsessive-Compulsive Disorder (OCD). They repeat behaviors that may have once made sense, but continually feel compelled to do so beyond any actual need. Our competitive society now actually selects for people who are OCD in this sense. They wind up in positions of power. See the chapter *The productivity of dead people* [Ch. 11] and *No Leaders* [Ch. 22].

We present our vision for what everyday life will look like post-scarcity, in the form of a fictional account *A day in the post-scarcity life* [Ch. 18].

Q: Why do you say scarcity encourages competition? If food is scarce, isn't it even more important that members of a tribe cooperate on hunting?

A: When properly executed, cooperation will produce more wealth per effort than competition because competition can only be as good as the best side minus the effort to fight, whereas cooperation can be better than the sum of the sides. In times of scarcity, additional wealth is all the more important. In the chapter, *Survival of the most cooperative* [Ch. 3], we discuss the tribal hunting example in detail. Yes, scarcity can increase the importance of the benefits of cooperating in the hunt. But scarcity may also increase the probability that others might, out of desperation, refuse to fairly share the spoils of the hunt.

Plus, there can be feedback loops. Competition encourages scarcity since it uses up resources, then there's more competition over the dregs, etc. Abundance encourages cooperation, producing more abundance, in a positive feedback loop.

Q: Without monetary incentives, things won't get done.

A: If things need doing, and nobody wants to do them, then we'll invent robots to do them. Increasing automation is what we've always done to improve productivity, and that trend is accelerating.

More generally, though, we don't like the idea of "incentive" as a motivation. See the chapter *Intrinsic and Extrinsic Motivation* [Ch. 10], particularly the section *The Bankruptcy of Incentive*. An incentive is some sort of bribe to get you to do something you don't otherwise want to do. As such, it is exploitive.

Instead, we'd like people to ask themselves the question, "Is it worth doing?" and "Do I want to do it?". If so, they'll do it. If not, and it's still worth doing, automate it.

Q: Will robots take all our jobs? AI and robots can never replace people in a lot of jobs.

A: Yes, robots will take our jobs and this is a good thing.

We don't want to get into an argument about whether every job can be automated, but the trend, likely to continue, is that a greater and greater percentage of jobs can be. So we better get ready for a future where a significant portion of the population will not have jobs. Agriculture once "employed" almost all working-age people, but now only 2% of US workers are in that industry [World Bank 2017].

The whole idea of a "job" is a creation of the Industrial Age; Capitalism and Communism were the economic technologies of that era. They'll both die as we move further into the Information Age and the Maker Age comes online. We can set up society so that people who are automated out of jobs need not suffer. And as Makerism takes hold, people will be able to make or do for themselves much of what their jobs' salary would buy. See *The productivity of dead people* [Ch. 11] and *Makerism* [Ch. 15].

Q: What will people do all day if they don't have jobs?

A: They'll do an incredible variety of things, once they don't have the pressure and structure of a single, standard "job". Some people who enjoy the job they have now, will continue to do much the same activity, only without the formal structure. People will make or consume music or art, talk to friends, invent, share, make what they need, raise barns with neighbors, teach, learn, wander around virtual realities, and things we haven't thought of yet.

There's of course the danger that some will just do drugs, watch TV, or settle into some other kind of dysfunctional, passive life, as some do now. The fact that there's economic pressure to have a job doesn't prevent this from happening. But we think most people want some sort of active and meaningful life, jobs or no jobs, as long as they can meet their material needs.

Another way to think of it is it's like what we now call "retirement". After age 65, you've "paid your debt to society" and many can retire on some sort of public or private stipend. Right now, a challenge for retirees is the social stigma of "obsolescence" associated with it in our culture, but that can change. Present-day economists think we need to raise the retirement age to keep Social Security solvent, but with Makerism, we can reduce the retirement age to 0. See *The productivity of dead people* [Ch. 11].

Q: There are a lot of details about the economy that you guys haven't figured out.

A: There are a lot of details we didn't cover, for brevity. Here's a few to give you the flavor.

Q: How will healthcare work?

A: Most healthcare will be done by people for themselves, as print-at-home tools and medicines will make personalized health care much cheaper and easier for laypeople. The smartphone, with cheap attachments, is leading the way in diagnostics. Big data collection will help individualize health care, making treatments more appropriate for exactly your conditions. Those motivated to help people by becoming doctors will continue to do so, and will be respected by people for doing it.

Q: Money and finance?

A: There might still be money, but its role will be steadily reduced as you can make more and more of your necessities. Why borrow for a house when you can just make its components and put it together? If you can make it again when it burns down, you don't need to insure it.

Q: Will we still have retail stores?

A: Stores become unnecessary. Many still like some aspects of the experience of shopping, as it affords social contact with other shoppers and merchandise experts, and physical encounters with products. Maybe we'll have "theme parks" that reproduce the positive aspects of this experience, without the enormous resource consumption of today's retail infrastructure. Some people might enjoy playing the role of salesperson, and give away their expertise to whoever will listen. Being heard is a strong motivator.

Q: You don't think the Capitalist elite will just stand by and let all this happen, do you? Once it starts to really eat into their power, they'll fight back.

A: Once it gets rolling, there's probably not much they can do to stop it. Like the Internet, the transformation we're predicting will be initiated by makers and other innovators for themselves and their communities, then it will spread. No company or government designed or wanted the Internet before it became popular (though initial research was government funded). We're hoping that this book will inspire you to be one of the innovators.

The powers that be will probably ignore most of it while it's developing, since it won't be operating in the circles of power and big money that they consider significant. We wish we could convince government and large companies to embrace new technologies for cooperation and manufacturing, as they could both profit from them in the short term, and help us get on the road to Makerism. Long term, well, the rich will still be rich, though they'll have more company. They'll be more likely to enjoy their Ferrari if there's less chance they'll be mugged in the parking lot.

How dare you trash democracy and voting?! That's what makes America great!

Q: Aren't the only choices democracy, or some sort of dictatorship?

A: Nope. We try to innovate about all sorts of things in our society, but for some reason people don't think of governments as something to be designed. Our education systems reinforce the idea that the only choices are the governments of the past.

We present one new alternative, which we call *Reasonocracy* [Ch. 25]. Of course we don't like dictatorships. But the actual mechanisms of today's US democracy are based on power relationships (voting and campaign finance), just as dictatorships are. Political discourse is competitive and contentious, causing gridlock. Cooperative reasoning can foster creative problem solving without the endless fights for power.

Q: Without leaders and hierarchies, things won't get done.

A: We think leaders and hierarchies actually stand in the way of getting things done. Today's corporate and government leadership hierarchies are basically a holdover from feudalism (*The beginning of history* [Ch. 19] and *No Leaders* [Ch. 22]). We now call them CEOs and Senators instead of Dukes and Earls, but same idea. We explore what really motivates people in (*Intrinisic and Extrinsic Motivation* [Ch. 10]). Some communities already operate without hierarchies. "Flattening the organization" is now an in-vogue business buzz-term.

Q: How can we possibly make decisions without voting?

A: Voting can be useful as a last resort when people have intractable differences. But why should we start out assuming every issue is intractable?

We look at consensus process (*Can Some of Us Get Along?* [Ch. 8] and *Some days in the life of a Reasonocrat* [Ch. 26]. *Reasonocarcy* is a structured problem solving approach that can usually reach far more optimal decisions than the all-too-common voting for the lesser of two evils.

We're inspired by the scientific community (*The process of Science* [Ch. 24]), which regularly provides constructive solutions to problems of great magnitude. The actual operation of science is mostly independent of leadership hierarchies, and there's very little voting. There's no President or CEO of Science, no political parties, no scientific courts, etc.

Q: Randomly select people to be representatives?! Some people will just turn out to be idiots.

A: As opposed to the congressmen selected by elections we have in the US? Congress has very low approval ratings, so a new way of selecting representatives has a low bar to jump over. Sure, the IQ of an average member of Congress is above average, but the motivations of those elected aren't aligned with the citizens so this extra IQ is used against citizens, not for them.

We model the representative selection process on jury duty, which recruits random people to make life-or-death decisions. But unlike juries, we propose minimum competence requirements. We propose educating representatives to be more open to ideas, better at cooperation and more skillful at reasoning. Furthermore, we don't throw them into the battlefield of power politics, but rather a structured process devoted to reasoning for optimal decision making.

Q, from the Right: Since you don't like Capitalism and you promote cooperation, aren't you Communists or Socialists?

A: No. Communist and Socialist societies have centralized planning. In *Makerism* [Ch. 15], everyone owns the means of production. There is no central planing for creating wealth. You make what you need. You don't make what you don't need unless you want to give it to a friend. The need for infrastructure and its central planning is greatly decreased.

The Right should love the increased freedom of choice, individualism, and smaller government of Makerism and Reasonocracy.

Q, from the Left: Technology is always a tool of the Capitalist Military-Industrial Complex. Aren't you just acting as their pawns?

A: No, but it's true that for centuries, many new technologies were originally developed and deployed for military applications. We believe that understanding cooperation and reducing scarcity will greatly reduce war. Since Makerism reduces or eliminates for-profit companies, the motivation of the Military-Industrial Complex evaporates. (See the *War* chapter [Ch. 33].)

The Left should love the promotion of peace, community building, and the reduction of poverty, inequality and discrimination, of Makerism and Reasonocracy.

Where's the bibliography and index?

It is publishing tradition (at the time of this writing) for nonfiction books to end with a bibliography and with an index. In hardcopy editions of this book, you may have arrived at this page by flipping through the book looking for them. Why aren't they here?

In today's world, some readers prefer reading online versions and some readers prefer the physical feeling of reading printed paper. We were mindful that many readers perceive very long printed books as intimidating (as well as heavy to carry around).

So we tried to keep it as concise as possible. Adding a full bibliography and index would have likely inflated the size of the book by at least 50 pages.

We believe that the functions of a bibliography and index are best provided by online tools. We don't have an index, but the functionality is satisfied by text search facilities available in reading software.

We have provided a full bibliography on our site,

http://www.whycantwe.org/bibliography

There are about 300 citations to external references in the book. We have taken care to provide a URL for every citation (inevitably, some will be out of date). You can search for the citation, (e.g., search for the text "[Fry 2011]", without the quotes) in the bibliography file. Clicking on that citation will take you to the original source material, if we could find it, or to a means of obtaining that source (such as a bookstore).

About the illustrations

Walt Lieberman

I am a glass artist by trade and the brother of Henry Lieberman, one of the authors. When Henry asked me to illustrate this book I was honored and excited. In recent years I have been a tour guide at the Museum of Glass in Tacoma. As part of that job I would often make drawings in chalk on the floor of the glassblowing studio to explain to museum visitors what the artists were making and the glassmaking processes. I became really interested in this rough technique. Drawing on a concrete floor with sidewalk chalks is not the most refined way of making art. It is, however fun and expressive. The ultimate purpose of these drawings is to explain complicated things more clearly, especially to folks who may be new to this subject. In that light, illustrating this book serves the same purpose. I want these drawings to help the reader follow the process of the authors' thinking. All the drawings for this book were done on the floor of the Museum of Glass Hot Shop floor in chalk.

My thanks to the Museum of Glass for the use of their floor and their support for my drawings.

http://www.facebook.com/walter.lieberman/

http://www.media.mit.edu/~lieber/Walt/

About the authors

Christopher Fry

Fry moved to Boston in 1973 to attend Berklee College of Music (the MIT of Jazz). Realizing his musical skills needed augmentation, he moved across the river to MIT (The Berklee of Computers). He's worked at BBN, IBM, MIT's Experimental Music Studio, Sloan (business) school & Media Lab and a host of start-ups. He's written languages for music composition (Computer Improvisation and Flavors Band), general purpose computing (Macintosh Common Lisp and Water) and decision support via reasoning (Justify). His latest language and development environment is to help makers describe processes for robots to make anything.

In 2012, Fry presented his plan for cutting gasoline use and the accident death rate (30K per year in the USA alone) in half (while saving money) to the Massachusetts Department of Transportation, the US House of Representatives, and the United Nations. Their disregard for this plan and the subsequent deaths of 200K+ people in auto accidents in the USA since, exposes these organizations' disregard for your life. This state of affairs provided motivation for this book.

Henry Lieberman

Henry Lieberman has been a manufacturer of fine intellectual property for the last 40 years. He is Research Scientist at the MIT Computer Science and Artificial Intelligence Lab (CSAIL), and has been Principal Research Scientist at the MIT Media Lab, running the Software Agents Group.

He is trying to make computers less stupid and frustrating than they are now, through the fields of Artificial Intelligence (AI) and Human-Computer Interaction (HCI). He has contributed to information overload by authoring more than 120 scientific papers and three books (before this one).

His research has impacted software on nearly every computer in the world. Nerds will be impressed that he invented real-time memory management, which made today's dynamic memory programming languages like Java, Python, and others possible. He is known for originating prototype object systems, found today in languages like Javascript. He has also been a pioneer in topics like programming by example and commonsense reasoning in AI.

He has a BS in mathematics from MIT and an HDR (PhD-equivalent) from the University of Paris (Sorbonne), where he was also a Visiting Professor. He has long collaborated with Fry on programming environments, debugging, AI, and several startups.

www.ingramcontent.com/pod-product-compliance
Lightning Source LLC
Chambersburg PA
CBHW061621220326
41598CB00026BA/3839